LONDON MATHEMATICAL SOCIETY LECTURE NOTE SERIES

Managing Editor: Professor J.W.S. Cassels
Department of Pure Mathematics and Mathematical Statistics
University of Cambridge, 16 Mill Lane
Cambridge CB2 1SB, England

The books in the series listed below are available from booksellers, or, in case of difficulty, from Cambridge University Press.

34 Representation theory of Lie groups, M.F. AITYAH *et al*
36 Homological group theory, C.T.C. WALL (ed)
39 Affine sets and affine groups, D.G. NORTHCOTT
40 Introduction to H_p spaces, P.J. KOOSIS
46 p-adic analysis: a short course on recent work, N. KOBLITZ
49 Finite geometrics and designs, P. CAMERON, J.W.P. HIRSCHFELD & D.R. HUGHES (eds)
50 Commutator calculus and groups of homotopy classes, H.J. BAUES
57 Techniques of geometric topology, R.A. FENN
59 Applicable differential geometry, M. CRAMPIN & F.A.E. PIRANI
62 Economics for mathematicians, J.W.S. CASSELS
66 Several complex variables and complex manifolds II, M.J. FIELD
69 Representation theory, I.M. GELFAND *et al*
74 Symmetric designs: an algebraic approach, E.S. LANDER
76 Spectral theory of linear differential operators and comparison algebras, H.O. CORDES
77 Isolated singular points of complete intersections, E.J.N. LOOIJENGA
78 A primer on Riemann surfaces, A.F. BEARDON
79 Probability, statistics and analysis, J.F.C. KINGMAN & G.E.H. REUTER (eds)
80 Introduction to the representation theory of compact and locally compact groups, A. ROBERT
81 Skew fields, P.K. DRAXL
82 Surveys in combinatorics, E.K. LLOYD (ed)
83 Homogeneous structures on Riemannian manifolds, F. TRICERRI & L. VANHECKE
86 Topological topics, I.M. JAMES (ed)
87 Surveys in set theory, A.R.D. MATHIAS (ed)
88 FPF ring theory, C. FAITH & S. PAGE
89 An F-space sampler, N.J. KALTON, N.T. PECK & J.W. ROBERTS
90 Polytopes and symmetry, S.A. ROBERTSON
91 Classgroups of group rings, M.J. TAYLOR
92 Representation of rings over skew fields, A.H. SCHOFIELD
93 Aspects of topology, I.M. JAMES & E.H. KRONHEIMER (eds)
94 Representations of general linear groups, G.D. JAMES
95 Low-dimensional topology 1982, R.A. FENN (ed)
96 Diophantine equations over function fields, R.C. MASON
97 Varieties of constructive mathematics, D.S. BRIDGES & F. RICHMAN
98 Localization in Noetherian rings, A.V. JATEGAONKAR
99 Methods of differential geometry in algebraic topology, M. KAROUBI & C. LERUSTE
100 Stopping time techniques for analysts and probabilists, L. EGGHE
101 Groups and geometry, ROGER C. LYNDON
103 Surveys in combinatorics 1985, I. ANDERSON (ed)
104 Elliptic structures on 3-manifolds, C.B. THOMAS
105 A local spectral theory for closed operators, I. ERDELYI & WANG SHENGWANG
106 Syzygies, E.G. EVANS & P. GRIFFITH
107 Compactification of Siegel moduli schemes, C-L. CHAI
108 Some topics in graph theory, H.P. YAP
109 Diophantine analysis, J. LOXTON & A. VAN DER POORTEN (eds)
110 An introduction to surreal numbers, H. GONSHOR
111 Analytical and geometric aspects of hyperbolic space, D.B.A. EPSTEIN (ed)
113 Lectures on the asymptotic theory of ideals, D. REES
114 Lectures on Bochner-Riesz means, K.M. DAVIS & Y-C. CHANG
115 An introduction to independence for analysts, H.G. DALES & W.H. WOODIN
116 Representations of algebras, P.J. WEBB (ed)
117 Homotopy theory, E. REES & J.D.S. JONES (eds)
118 Skew linear groups, M. SHIRVANI & B. WEHRFRITZ
119 Triangulated categories in the representation theory of finite-dimensional algebras, D. HAPPEL
121 Proceedings of *Groups – St. Andrews 1985*, E. ROBERTSON & C. CAMPBELL (eds)
122 Non-classical continuum mechanics, R.J. KNOPS & A.A. LACEY (eds)
124 Lie groupoids and Lie algebroids in differential geometry, K. MACKENZIE
125 Commutator theory for congruence modular varieties, R. FREESE & R. MCKENZIE

126 Van der Corput's method of exponential sums, S.W. GRAHAM & G. KOLESNIK
127 New directions in dynamical systems, T.J. BEDFORD & J.W. SWIFT (eds)
128 Descriptive set theory and the structure of sets of uniqueness, A.S. KECHRIS & A. LOUVEAU
129 The subgroup structure of the finite classical groups, P.B. KLEIDMAN & M.W. LIEBECK
130 Model theory and modules, M. PREST
131 Algebraic, extremal & metric combinatories, M-M. DEZA, P. FRANKL & I.G. ROSENBERG (eds)
132 Whitehead groups of finite groups, ROBERT OLIVER
133 Linear algebraic monoids, MOHAN S. PUTCHA
134 Number theory and dynamical systems, M. DODSON & J. VICKERS (eds)
135 Operator algebras and applications, 1, D. EVANS & M. TAKESAKI (eds)
136 Operator algebras and applications, 2, D. EVANS & M. TAKESAKI (eds)
137 Analysis at Urbana, I, E. BERKSON, T. PECK & J. UHL (eds)
138 Analysis at Urbana, II, E. BERKSON, T. PECK & J. UHL (eds)
139 Advances in homotopy theory, S. SALAMON, B. STEER & W. SUTHERLAND (eds)
140 Geometric aspects of Banach spaces, E.M. PEINADOR and A. RODES (eds)
141 Surveys in combinatorics 1989, J. SIEMONS (ed)
142 The geometry of jet bundles, D.J. SAUNDERS
143 The ergodic theory of discrete groups, PETER J. NICHOLLS
144 Introduction to uniform spaces, I.M. JAMES
145 Homological questions in local algebra, JAN R. STROOKER
146 Cohen-Macaulay modules over Cohen-Macaulay rings, Y. YOSHINO
147 Continuous and discrete modules, S.H. MOHAMED & B.J. MÜLLER
148 Helices and vector bundles, A.N. RUDAKOV et al
149 Solitons, nonlinear evolution equations and inverse scattering, M.A. ABLOWITZ & P.A. CLARKSON
150 Geometry of low-dimensional manifolds 1, S. DONALDSON & C.B. THOMAS (eds)
151 Geometry of low-dimensional manifolds 2, S. DONALDSON & C.B. THOMAS (eds)
152 Oligomorphic permutation groups, P. CAMERON
153 L-functions and arithmetic, J. COATES & M.J. TAYLOR (eds)
154 Number theory and cryptography, J. LOXTON (ed)
155 Classification theories of polarized varieties, TAKAO FUJITA
156 Twistors in mathematics and physics, T.N. BAILEY & R.J. BASTON (eds)
157 Analytic pro-*p* groups, J.D. DIXON, M.P.F. DU SAUTOY, A. MANN & D. SEGAL
158 Geometry of Banach spaces, P.F.X. MÜLLER & W. SCHACHERMAYER (eds)
159 Groups St Andrews 1989 Volume 1, C.M. CAMPBELL & E.F. ROBERTSON (eds)
160 Groups St Andrews 1989 Volume 2, C.M. CAMPBELL & E.F. ROBERTSON (eds)
161 Lectures on block theory, BURKHARD KÜLSHAMMER
162 Harmonic analysis and representation theory for groups acting on homogeneous trees, A. FIGA-TALAMANCA & C. NEBBIA

London Mathematical Society Lecture Note Series. 164

Quasi-Symmetric Designs

Mohan S. Shrikhande
Department of Mathematics, Central Michigan University

Sharad S. Sane
Department of Mathematics, University of Bombay

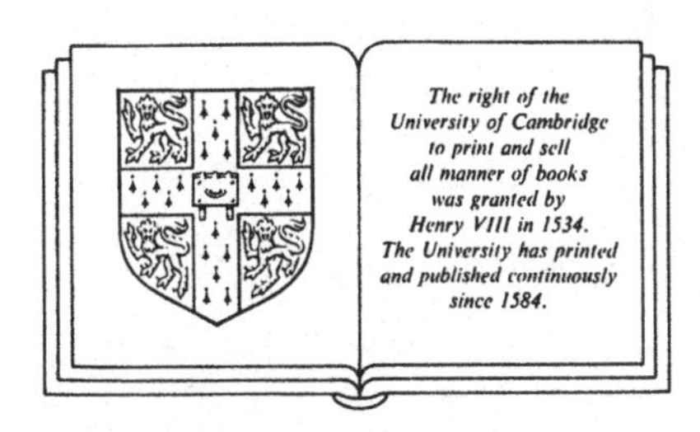

CAMBRIDGE UNIVERSITY PRESS

Cambridge
New York Port Chester
Melbourne Sydney

CAMBRIDGE UNIVERSITY PRESS
Cambridge, New York, Melbourne, Madrid, Cape Town, Singapore, São Paulo

Cambridge University Press
The Edinburgh Building, Cambridge CB2 8RU, UK

Published in the United States of America by Cambridge University Press, New York

www.cambridge.org
Information on this title: www.cambridge.org/9780521414074

First published 1991

A catalogue record for this publication is available from the British Library

Library of Congress Cataloguing in Publication data

Shrikhande, Mohan S., 1945–
Quasi-symmetric designs / Mohan S. Shrikhande, Sharad S. Sane.
p. cm. – (London Mathematical Society lecture note series; 164)
Includes bibliographical references and index.
ISBN 0-521-41407-5 (U.S. : pbk.)
1. Block designs. 2. Graph theory. 3. Coding theory. I. Sane,
Sharad S., 1950– . II. Title. III. Series.
QA166.3.S57 1991
519.5—dc20 91-762 CIP

ISBN 978-0-521-41407-4 paperback

Transferred to digital printing 2007

To Neelima and Aditi, from Mohan

To B.V. Limaye, from Sharad

CONTENTS

	Preface	ix
I.	Basic results from designs	1
II.	Strongly regular graphs and partial geometries	17
III.	Basic results on quasi-symmetric designs	34
IV.	Some configurations related to strongly regular graphs and quasi-symmetric designs	49
V.	Strongly regular graphs with strongly regular decompositions	82
VI.	The Witt designs	99
VII.	Extensions of symmetric designs	122
VIII.	Quasi-symmetric 2-designs	141
IX.	Towards a classification of quasi-symmetric 3-designs	174
X.	Codes and quasi-symmetric designs	192
	References	207
	Index	222

PREFACE

Combinatorics is generally concerned with counting arrangements within a finite set. One of the basic problems is to determine the number of possible configurations of a given kind. Even when the rules specifying the configuration are relatively simple, the questions of existence and enumeration often present great difficulties. Besides counting, combinatorics is also concerned with questions involving symmetries, regularity properties, and morphisms of these arrangements. The theory of block designs is an important area where these facts are very apparent. The study of block designs combines number theory, abstract algebra, geometry, and many other mathematical tools including intuition. In the words of G.C. Rota (in: Studies in Combinatorics, Mathematical Association of America, 1978),

" Block designs are generally acknowledged to be the most complex mathematical structures that can be defined from scratch in a few lines. Progress in understanding and classification has been slow and proceeded by leaps and bounds, one ray of sunlight followed by years of darkness. ...This field has been enriched and made even more mysterious, a battleground of number theory, projective geometry and plain cleverness. This is probably the most difficult combinatorics going on today..."

In the last few years, some new text-books (Beth, Jungnickel and Lenz, Hughes and Piper, Wallis) on Design Theory have been published. Dembowski's 'Finite Geometries', M. Hall Jr.'s 'Combinatorial Theory,' and Ryser's 'Combinatorial Mathematics' are regarded as some of the classic references in combinatorics, particularly in the area of designs. Block designs have connections with group theory, graph theory, coding theory, and number theory. The monographs of Biggs and White; Cameron and van Lint are mostly devoted to the developments in specialized topics and show how progress in one of these areas impacts design theory. In spite of the appearance of some recent monographs (particularly the ones by Lander on "Symmetric Designs: An Algebraic Approach", Payne and Thas on "Generalized Quadrangles", Batten on "Combinatorics of Finite Geometries") we believe that not many monographs dealing with special topics in design theory are available at present. This is one of the motivations for writing the present monograph.

As is well known, non-trivial designs satisfy the inequality $v \leq b$, i.e., the number of blocks is at least as large as the number of points. The situation of equality is called a symmetric design. This is an important class of designs which is characterized by the property that the design has precisely one (block) intersection number. Quasi-symmetric designs are designs (i.e., 2-designs) with at most two intersection numbers. With this general definition, symmetric designs are just improper quasi-symmetric designs (q.s. designs). The theory of symmetric designs is mathematically enriched by results such as the Bruck-Ryser-Chowla theorem and Lander's monograph is devoted to this aspect of symmetric designs. No similar strong tools seem to be known for proper quasi-symmetric designs. There have, however, been recent attempts in this direction, mainly by Calderbank, using tools from coding theory. On the other hand, much investigation on (proper) q.s. designs is facilitated by the block graph associated with such a design: If x and y are the two intersection numbers, then make two vertices (blocks) adjacent if the corresponding blocks intersect in x points. This graph turns out to be a strongly regular graph and is non-trivial if the design is not symmetric. This early observation paved the way for an intersection between theories of quasi-symmetric designs and strongly regular graphs. The latter objects were defined by Bose and the most elegant examples are provided by groups with rank three permutation action. Though many expository articles on strongly regular graphs exist, no comprehensive book dealing with all the aspects of this topic seems to be available at present. This fact has also been mentioned in the very recent book on distance regular graphs by Brouwer, Cohen, and Neumaier.

At this stage, we discuss the organization of the chapters in this monograph. The material is divided into ten chapters, the first three of which cover basic material on designs, strongly regular graphs, and quasi-symmetric designs respectively. Various standard examples of symmetric designs (Hadamard designs and projective geometries in particular), strongly regular graphs, and quasi-symmetric designs (affine designs in particular) are included. Our treatment also includes the rationality conditions useful in pinning down the parameters of strongly regular graphs and quasi-symmetric designs.

Historically, the forerunners of strongly regular graphs are partially balanced designs and association schemes. Partially balanced designs are generalizations of designs in which the simultaneous occurrence of a point-pair in the blocks is determined by the superposed point-graph. The general question in this regard is that of the determination of suitable conditions under which a strongly regular graph gives rise to a strongly regular subgraph or a partially balanced design as a substructure. Two important classes of strongly regular graphs studied in Chapter IV are the Mesner family of graphs and the (block) graphs of certain generalized quadrangles. Included among other things are the Bose-Connor property of semiregular group divisible designs, results on special partially balanced designs, and an eigenvalue characterization of partial geometric designs. The theme developed in Chapter IV is further continued in Chapter V, where a detailed discussion of the recent Haemers-Higman study of strongly regular decompositions is made. Included in that chapter are the Hoffman and Cvetcovic coclique bounds and the interlacing theorems.

The construction problem of designs was first handled by statisticians since designs with the right parameters were needed in the design of experiments (see, for example, the book of Raghavarao on designs). However, designs with an aesthetic appeal and elegance of construction are obtained by construction from groups; the constructions of Witt designs from Mathieu groups, in our opinion, bear the best testimony to this fact. A combinatorial exposition of Witt designs was given by Lüneberg in the late 1960's and Chapter VI is devoted to the constructions and combinatorial properties of Witt designs. There are two other independent reasons for the inclusion of Witt designs in this monograph. First, these designs give examples of quasi-symmetric designs and strongly regular graphs (such as the Higman-Sims, Hoffman-Singleton and McLaughlin graphs). In this connection, it should also be pointed out that substantial work in design theory today concerns itself with various characterizations of Witt designs. Secondly, the Witt designs on 24 and 12 points give examples of Steiner systems S(t, k, v) with $t = 5$, and no non-trivial Steiner system with $t \geq 6$ seems to be known. Many expositions of Witt designs are available in the literature and a similar treatment was given by van Lint. Our treatment in Chapter VI is self-contained and can be understood with no knowledge of group theory.

While the symmetric designs by themselves are improper quasi-symmetric, the 3-designs obtained as extensions of symmetric designs are proper quasi-symmetric, with intersection number pair $(x, y) = (0, y)$. In fact, an observation of Cameron states that the condition $(x, y) = (0, y)$ for a q.s. 3-design characterizes it as an extension of a symmetric design. In his celebrated theorem, Cameron uses the preceeding statement to classify the parameters of all the symmetric designs that can be possibly extended. Much recent activity in the area of quasi-symmetric designs is an outcome of Cameron's theorem. Barring the infinite class of Hadamard 3-designs (whose existence is equivalent to the existence of a Hadamard matrix of the corresponding order and whose block graph is just a 1-factor of the complete graph) and two other sporadic examples (one of which is an extension of a projective plane of order ten shown not to exist by a result of Lam and others), the classification theorem of Cameron also includes a putative family of 3-designs, the first object of which is the Witt design on 22 points. In spite of a recent result of Bagchi, it seems difficult to determine whether or not these 3-designs actually exist (particularly since the derived designs themselves are unknown for $\lambda \geq 3$). Chapter VII deals with this infinite family of 3-designs and their residuals. Among other things, we include a proof to show that a symmetric design is a block residual if and only if its dual is also a block residual.

Chapter VIII is devoted to (general) quasi-symmetric 2-designs and is an important chapter. Observe that most of the 3-designs occurring in Cameron's extension theorem, mentioned in the previous paragraph, are actually triangle-free, i.e., have no three mutually disjoint blocks. Though the concept of a quasi-symmetric design dates back to the late 1960's, it was perhaps only in the early 1980's that the structural investigations of quasi-symmetric designs began with the introduction of a polynomial tool for the study of triangle-free q.s. designs (with $(x, y) = (0, y)$). A large part of Chapter VIII is devoted to results that exploit this polynomial approach. An outstanding conjecture of M. Hall Jr. states that for a fixed $\lambda \geq 2$ there are finitely many symmetric (v, k, λ)-designs. Among other results, ChapterVIII includes a proof of the following analogue of this conjecture: For a fixed $\lambda \geq 2$, there are finitely many proper q.s. (v, k, λ)-designs with the smaller intersection number 0.

Are there t-designs with $t \geq 3$ that are also quasi-symmetric? Cameron showed that non-triviality implies $t \leq 4$. Ito and others proved that up to complementaion the only q.s. 4-design is the Witt design $S(4, 7, 23)$. This theme is a particular situation of the Ray-Chaudhuri and Wilson result on tight t-designs. A quadratic with coefficients in the design parameters v, k and λ and whose zeros are the two intersection numbers x and y of a q.s. 3-design is implicit in the work of Delsarte. A somewhat explicit formulation of this polynomial has been particularly useful in recent investigations. Chapter IX includes a short proof of a recent result which shows that a non-trivial q.s. 3-design with the smaller intersection number one must be the S(4, 7, 23) or its residual. This chapter also includes a conjecture on q.s. 3-designs and an account of some results which indicate a support of that conjecture. If proved, the conjecture will give a result much stronger than that of Ito and others on q.s. 4-designs.

As we already remarked, the recent use of tools from other areas in the theory of quasi-symmetric designs has proved fruitful. The work of Calderbank and Tonchev uses coding theory and other branches and has succeeded in giving many non-existence results for q.s. designs. The last Chapter X is a brief description of these results. Included in Chapter VII is a table of possible parameters of q.s. 2-designs (with small parameters) with certain additional conditions prepared by Neumaier. Neumaier's table seems to have been a motivation for some non-existence results mentioned in the preceeding paragraph. We would, however, like to point out that exhaustive tables of q.s. designs will certainly be useful. No such elaborate tables are available at present.

While the last two decades have witnessed some interesting activity in the subject matter, there has not been a single reference dealing exclusively with the topics covered. Ours is an attempt to fill that void. We hope that this monograph will be useful not only to research workers but also to beginning graduate students who would like to gain some acquaintance with the area.

ACKNOWLEDGMENTS

This monograph is an outgrowth of a project spread over several years. In the academic year 1984-85 , SSS was at Central Michigan University as a Visiting Research Professor. During this period the authors collaborated on several problems on quasi-symmetric designs. In late 1987 they decided to write a monograph on this topic. MSS would like to acknowledge a Central Michigan University Research Professorship in the Fall of 1988, during which a major portion of the initial draft of this monograph was written. In the academic year 1989-90, SSS was again visiting Central Michigan University. During this time, he was able to work on the manuscript and both the authors completed the final version. SSS gratefully acknowledges the financial support and hospitality provided by Central Michigan University during his visits. He also thanks the University of Bombay for granting him leave for these visits. Finally, both of us would like to thank Barbara Curtiss at Central Michigan University for her splendid job of typing this material.

Mohan S. Shrikhande
Sharad S. Sane

Mt. Pleasant and Bombay
March 1991

I. BASIC RESULTS FROM DESIGNS

In this first chapter, we collect together and review some basic definitions, notation, and results from design theory. All of these are needed later on. Further details or proofs not given here may be found, for example, in Beth, Jungnickel and Lenz [15], Dembowski [61], Hall [74], Hughes and Piper [95] , or Wallis [177]. We mention also the monographs of Cameron and van Lint [49], Biggs and White [18], and the very recent one by Tonchev [175].

Let $X = \{x_1, x_2, \ldots, x_v\}$ be a finite set of elements called points or treatments and $\beta = \{B_1, B_2, \ldots, B_b\}$ be a finite family of distinct k-subsets of X called blocks. Then the pair $D = (X, \beta)$ is called a t-(v, k, λ) design if every t-subset of X occurs in exactly λ blocks. The integers v, k, and λ are called the parameters of the t-design D. The family consisting of all k-subsets of X forms a k-$(v, k, 1)$ design which is called a complete design. The trivial design is the v-$(v, v, 1)$ design. In order to exclude these degenerate cases we assume always that $v > k > t \geq 1$ and $\lambda \geq 1$. We use the term finite incidence structure to denote a pair (X, β), where X is a finite set and β is a finite family of not necessarily distinct subsets of X. In most of the situations of interest in the later chapters, however, we will have to tighten these restrictions further. For example, though we do not impose the condition that the blocks be distinct sets, that usually would be the case in view of some other stipulations.

A t-design, or more generally an incidence structure, is completely specified up to labellings of its points and blocks by its usual (0, 1)-incidence matrix N. This matrix $N = (n_{ij})_{v \times b}$ is defined by $n_{ij} = 1$ or 0 according as $x_i \in B_j$ or not. Two designs D_1 and D_2 are said to be isomorphic (denoted by $D_1 \cong D_2$) if there are bijections between their point-sets and block-sets respectively which preserve the incidence. Equivalently, the incidence matrix N_1 of D_1 can be changed to N_2 of D_2 by permuting rows and columns.

We give below three well known and small examples.

Example 1.1. The following picture is a 2-(7, 3, 1) design called the Fano plane. Here the blocks are triples of points which lie on a line or circle.

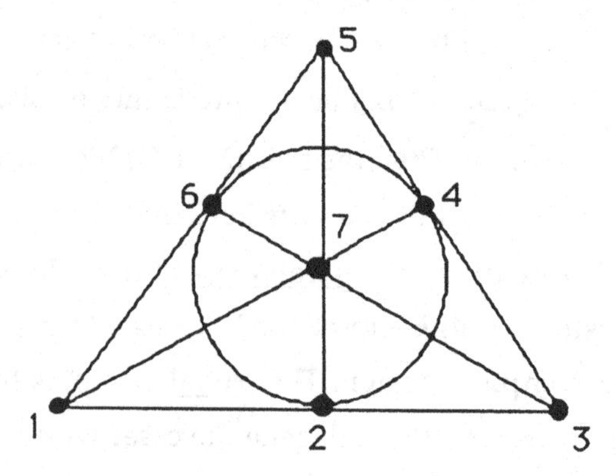

Example 1.2. Take complements of blocks in Example 1.1. We obtain a 2-(7, 4, 2) design.

Example 1.3. In Example 1.1, add a new symbol ∞ to the point set. Form new blocks by taking the complements of old blocks; in addition take old blocks adjoined with ∞. This new design is a 3-(8, 4, 1) which is an example of the "smallest Hadamard 3-design." The general construction is mentioned later on.

The following simple observation is an important tool in combinatorics. It is known as the method of two way counting and is so commonly used in design theory that we often use phrases such as "two way counting produces" or "counting **S** in two ways gives" etc.

Lemma 1.4. Let $\boldsymbol{U}$ and $\boldsymbol{V}$ be finite sets and let $\boldsymbol{S} \subseteq \boldsymbol{U} \times \boldsymbol{V} = \{(u,v) : u \in \boldsymbol{U}, v \in \boldsymbol{V}\}$. For all $a \in \boldsymbol{U}, b \in \boldsymbol{V}$, define subsets of $\boldsymbol{S}$ by

$$\boldsymbol{S}(a,.) = \{(u,v) \in \boldsymbol{S} : u = a\} \text{ and } \boldsymbol{S}(.,b) = \{(u,v) \in \boldsymbol{S} : v = b\}$$

Then $|\boldsymbol{S}| = \sum_{a \in \boldsymbol{U}} |\boldsymbol{S}(a,.)| = \sum_{b \in \boldsymbol{V}} |\boldsymbol{S}(.,b)|$.

As an immediate application of the above, we have the following result.

Theorem 1.5. Let $D = (X, \beta)$ be a t-(v, k, λ) design. Then the following assertions hold.

(a) For i = 0, 1, ..., t-1, if λ_i denotes the number of blocks containing i points, then λ_i is independent of the choice of i points and in fact, $\lambda_i = \lambda_{i+1} \frac{v-i}{k-i}$.

(b) D is also an i-(v, k, λ_i) design for i = 1, 2, ..., t-1, where

$$\lambda_i = \frac{(v-i)(v-i-1)\cdots(v-t+1)}{(k-i)(k-i-1)\cdots(k-t+1)}\lambda .$$

Proof. (b) follows from (a) using the formula for λ_i in terms of λ_{i+1} and use of induction. Consider (a). We make an induction on j = t-i, i = 0, 1,..., t-1. Assume that any set of i+1 = t-(j-1) points is contained in exactly λ_{i+1} blocks of D. Let $\{x_1, x_2, ..., x_i\}$ be some set of i points of D. Count pairs (x, B) in two ways, where $x \notin \{x_1, x_2, ..., x_i\}$ and $\{x_1, x_2, ..., x_i, x\}$ is contained in B. Then Lemma 1.4 gives $(v-i)\,\lambda_{i+1} = (k-i)\,\lambda_i$ which gives λ_i in terms of λ_{i+1} as desired and also shows that λ_i is an invariant of D. Observe that our argument has proved both the basis and induction step of our proof.

Remark 1.6. An obvious necessary condition for the existence of a t-(v, k, λ) design is that λ_i be integral for $0 \leq i \leq t$. This shows, for example, that a 3-(11, 4, 1) design does not exist.

We have in a t-(v, k, λ) design D that $\lambda_0 = b$, the number of blocks, and denote by $\lambda_1 = r$ the number of blocks through any point of D. A t-(v, k, λ) design is also denoted by $\mathbf{S}_\lambda(t, k, v)$. A Steiner system is an $\mathbf{S}_1(t, k, v)$. A 2-design (which is not trivial nor complete) is called a balanced incomplete block design (BIBD) or simply a design. For 2-designs, counting flags (i.e., incident point-block pairs) and then 2-flags (i.e., incident pairs of points with blocks) we have

Lemma 1.7. The parameters v, b, r, k, λ of a 2-design satisfy $bk = vr$ and $\lambda(v\text{-}1) = r(k - 1)$.

Remark 1.8. Note that the first relation in Lemma 1.7 holds for any 1-design and both the relations are also immediate consequences of Theorem 1.5 (a) with i = 0 and 1 respectively. Since it sometimes serves our purpose better to list all the parameters, a (v, b, r, k, λ) design is simply a (v, k, λ) design where r and b are given by Lemma 1.7 and must be integers. Then the obvious necessary conditions for the existence of a 2-(v, k, λ) design are $\lambda(v\text{-}1) \equiv 0 \pmod{k\text{-}1}$ and $\lambda v(v\text{-}1) \equiv 0 \pmod{k(k\text{-}1)}$. Hanani [76] has shown these conditions to be sufficient for k = 3, 4, and 5, and every λ, with the exception $v = 15$, $k = 5$, $\lambda = 2$. Wilson [181] has proved the necessary conditions to be sufficient asymptotically. Recently, Teirlinck [168] has proved that non-trivial t-designs without repeated blocks exist for all t. This was a major unsolved problem in design theory.

Our discussion above should make it sufficiently clear that the existence problem of 2-designs is much harder than the one for 1-designs in view of the following result.

Theorem 1.9. A 1-design (v, b, r, k) without repeated blocks exists if and only if vr=bk and b $\leq \binom{v}{k}$.

We give below D. Billington's [19] elegant proof of this result.

Proof. Observe that a 1-design is just a special case of a $(k, r_1, r_2, ..., r_v)$ - design which consists of a collection of b blocks of cardinality k chosen from a v-set X={1, 2, ..., v}, such that no two blocks are the same, and each $i \in X$ occurs exactly r_i times (i=1, 2, ..., v). A 1-design satisfies $r_1 = r_2 = ... = r_v = r$.

<u>Assertion</u>: Suppose a $(k, r_1, r_2, ..., r_v)$-design exists. If $r_i > r_j$ for some $i \neq j$, then a $(k, r_1, r_2, ..., r_{i-1}, r_i-1, r_{i+1}, ..., r_{j-1}, r_j+1, r_{j+1}, ..., r_v)$-design exists.

To prove the assertion, suppose D is a $(k, r_1, r_2, ..., r_v)$-design with $r_i > r_j$ for some $i \neq j$. Let $B_1, B_2, ..., B_n$ be all the blocks which contain i, and $C_1, C_2, ..., C_m$ be all the blocks which contain j. Since $r_i > r_j$, we necessarily have $0 \leq m < n$. Hence D has a block $B \in \{B_1, B_2, ..., B_n\}$ such that $B \notin \{C_1, C_2, ..., C_m\}$. Form $B^* = (B \setminus \{i\}) \cup \{j\}$. Omitting block B from D and replacing it by B^* results in a $(k, r_1, r_2, ..., r_{i-1}, r_i-1, r_{i+1}, ..., r_{j-1}, r_j+1, r_{j+1}, ..., r_v)$-design. This proves the assertion.

We return to the proof of the theorem. The necessity is obvious. Suppose now vr = bk and b $\leq \binom{v}{k}$. Choose any collection of b distinct k-sets from X = {1, 2, ..., v} obtaining a $(k, r_1, r_2, ..., r_v)$-design D for which $vr = bk = r_1 + r_2 + ... + r_v$. That is, r is the average of $r_1, r_2, ..., r_v$. If $r_i > r_j$ for some $i \neq j$, then using the assertion, we can obtain a $(k, r_1, ..., r_{i-1}, r_i-1, r_i+1, ..., r_{j-1}, r_j+1, ..., r_v)$-design D' for which the average of the replication numbers is still r. Successive use of the assertion results in a (k, r, r, ..., r)-design. This completes the proof of the theorem.

The next result is easily proved.

Lemma 1.10. The incidence matrix N of a 2-(v, k, λ) design satisfies

(i) $NN^t = (r - \lambda)I + \lambda J$, where I is the identity matrix of order v and J the all one matrix.

(ii) $\det(NN^t) = (r - \lambda)^{v-1} rk$.

We now can get Fisher's inequality for 2-designs.

Theorem 1.11. In a non-trivial 2-(v, k, λ) design, the number of blocks b is greater than or equal to the number of points v.

Proof. Here $\det(NN^t) \neq 0$. Thus $v = \text{rank}(NN^t) \leq \text{rank}(N) \leq b$.

A 2-(v, k, λ) design is called symmetric if $v = b$. The next result gives various characterizations of symmetric designs.

Theorem 1.12. The following are equivalent for a 2-(v, k, λ) design D.

(i) D is symmetric.

(ii) $r = k$.

(iii) Any two blocks intersect in λ points.

(iv) The dual D^t of D is a 2-design.

Proof. (Outline) From the proof of Theorem 1.11, $v = b$ implies that N is non-singular and by Lemma 1.7, r = k. Since $NN^t = (r-\lambda)I + \lambda J = (k-\lambda)I + \lambda J$ and $NJ = kJ$ we have $N^{-1}J = k^{-1}J$, which can be substituted in $N^tN = N^{-1}NN^tN$ to obtain $N^tN = (k-\lambda)I + \lambda J = NN^t$. This proves that (i) and (ii) are equivalent and imply (iii), which clearly implies (iv). If (iv) holds, then Theorem 1.10 applied to D^t gives $b \leq v$ and $v \leq b$. So (i) holds.

We recall that the dual D^t of a design D is obtained by interchanging the roles of points and blocks. Furthermore, a point B_j is incident with block x_i in D^t if x_i is incident with B_j in D.

Exercise 1.13. Give a matrix-free proof of Theorems 1.10 and 1.11. In fact the following stronger result (Beth et al. [15] or Tonchev [175]) can be proved.

Theorem 1.14. If D is a design with s repeated blocks, then $b \geq sv$.

From Lemma 1.10, if D is symmetric then $(\det(N))^2 = \det(NN^t) = (k-\lambda)^{v-1}k^2$. Since all the numbers involved are integers, we obtain the following result proved independently by Schutzenberger [142], Chowla and Ryser [55], and S.S. Shrikhande [157].

Theorem 1.15. If there exists a symmetric 2-(v, k, λ) design with v even, then (k - λ) must be a square.

The above theorem rules out, for example, a symmetric 2-(22, 7, 2) design. The case of odd v (Theorem 1.16) needs deeper number theoretic arguments than the simple matrix proof of Theorem 1.15. Theorem 1.16 was proved by Chowla and Ryser [55] (a proof based on the calculation of the Hasse-Minkowski invariant was also given by

S.S. Shrikhande [157]) and Theorems 1.15 and 1.16 are together now known by the name Bruck-Ryser-Chowla Theorem.

Theorem 1.16. A necessary condition for the existence of a symmetric 2-(v, k, λ) design with odd v is that the equation $x^2 = (k - \lambda)\, y^2 + (-1)^{(v-1)/2}\lambda z^2$ has an integral solution $(x, y, z) \neq (0, 0, 0)$.

The above result rules out, for example, a symmetric 2-(43, 7, 1) design. It was not known whether the conditions of the Bruck-Ryser-Chowla Theorem were sufficient. Apparently they are not, in view of a recent paper by Lam et al. [102].

Given a symmetric 2-(v, k, λ) design D and a block B, we can obtain a 2-design D^B called the residual of D with respect to B. The points of D^B are the points of D outside B. The blocks of D^B are the blocks of D minus the points of B. It is easily checked that D^B is a 2-$(v\text{-}k, k\text{-}\lambda, \lambda)$ design. A 2-design is quasi-residual if it has the right parameters to be the residual of a suitable symmetric 2-design. To be specific a 2-(v, k, λ) design is called quasi-residual if $r = k + \lambda$ (equivalently $v = k(k + \lambda - 1)/\lambda$).

For $\lambda = 1$, any quasi-residual design D is a 2-$(k^2, k, 1)$ design, i.e., an affine plane. The familiar process of embedding an affine plane in a projective plane immediately shows that a quasi-residual design with $\lambda = 1$ is the residual of a unique symmetric 2-design. The following result of Hall and Connor [75] covers the case $\lambda = 2$.

Theorem 1.17. Any quasi-residual 2-$(v, k, 2)$ design is the residual of a unique symmetric 2-design.

We should also mention that the exact analogue of Theorem 1.17 for $\lambda \geq 3$ is not true in general and there exist counterexamples (e.g. [74]). However Bose, S.S. Shrikhande, and Singhi [30] proved that there exists a function $f(\lambda)$ on the positive integers such that any quasi-residual 2-(v, k, λ) design with $\lambda \geq 3$ is the residual of a unique symmetric 2-design if $k > f(\lambda)$.

Let D be a t-(v, k, λ) design and p a point of D. The derived design (or the point contraction) D_p with respect to p is the $(t - 1)$-$(v - 1, k - 1, \lambda)$ design whose points are the points of D other than p and whose blocks are the blocks of D passing through p. The residual design D^p is a $(t - 1)$- $(v - 1, k, \lambda_{t-1} - \lambda)$ design, whose point set is that of D_p and whose blocks are those blocks of D missing p. A t-design D is called extendable if there exists a $(t+1)$- design D^* and a point ∞ of D^* such that $D \cong D^*_\infty$. If D is a given t-(v, k, λ) design, then all the derived designs of D are $(t-1)$-$(v-1, k-1, \lambda)$ designs and therefore have identical parameter sets. This simple observation gives as a special case

Lemma 1.18. If D is a 3-design and if for some point p the derived design D_p is a symmetric design, then every derived design D_q of D is a symmetric design with the same parameters as those of D_p.

Proof. Apply the definition of a symmetric design and observe that D_p and D_q have identical parameters.

The naive looking Lemma 1.18 is an important tool in the proof of Cameron's theorem discussed at the end of this chapter. The next result of Hughes [92] gives a simple necessary condition for extendability.

Proposition 1.19. If a t-(v, k, λ) design with b blocks is extendable, then $k + 1$ divides $b(v + 1)$.

Proof. Let D be a t-(v, k, λ) design and D^* its extension. Apply the condition $bk = vr$ to D^* and note that here the number of blocks through a point of D^* is the number of blocks of D.

Applying Proposition 1.19 to a projective plane of order n (= symmetric 2-$(n^2 + n + 1, n + 1, 1)$ design), Hughes [92] obtained the following:

Theorem 1.20. If a projective plane of order n is extendable, then $n = 2, 4$, or 10.

Remarks 1.21. The projective plane of order 2 is uniquely extendable to a 3-(8, 4, 1) design. The projective plane of order 4 is three times extendable, giving the famous and unique 3-(22, 6, 1), 4-(23, 7, 1), and 5-(24, 8, 1) designs (see e.g. [15]). Lam et al. [101], using a computer, have shown that a plane of order ten is not extendable. (Lam, Thiel, and Swiercz have apparently, very recently, ruled out the existence of a plane of order ten [102]).

We now recall some well known and important constructions for designs. Some of these will be often referred to later on. We list these facts as remarks, and for details refer to the references mentioned earlier.

Remarks 1.22. A Hadamard matrix H is an $n \times n$ matrix with entries ± 1 which satisfies $HH^t = H^tH = nI$. It is clear that multiplying the rows or columns of a Hadamard matrix by ± 1 produces another Hadamard matrix. The matrices [1] and $\begin{bmatrix} 1 & 1 \\ -1 & 1 \end{bmatrix}$ are Hadamard matrices of orders 1 and 2. It can be shown that if there exists a Hadamard matrix of order $n > 2$, then $n \equiv 0 \pmod 4$. Whether the converse holds is an outstanding open problem. There are several methods of constructing Hadamard matrices for which the reader may consult the references mentioned earlier. The smallest order of a Hadamard matrix which is in doubt is 428.

Remarks 1.23. Let H be a Hadamard matrix of order $n \geq 8$. By suitably multiplying rows and columns of H by -1, we may assume that the first row and column of H are all 1's. Now delete this row and column and replace all occurrences of -1 by 0. In this manner we obtain a matrix N which is the incidence matrix of a symmetric 2-$(n - 1, (n/2) - 1, (n/4) - 1)$-design. Such designs are known as Hadamard 2-designs. Conversely, from a symmetric 2-design with the above parameters we can recover a Hadamard matrix. As examples of Hadamard 2-designs, we mention the Paley designs, with n-1 = q, a prime or prime power, where $q \equiv 3 \pmod 4$. The points of the design are elements of GF(q), and the blocks are sets of the form $Q + a$ ($a \in$ GF(q)), where Q is the set of all non-zero squares in GF(q). The Fano plane in Example 1.1 is the case where $q = 7$.

Remarks 1.24. Let H be a Hadamard matrix of order $n \geq 8$ having all 1's in the first row. Then every row other than the first has $n/2$ entries 1 and $n/2$ entries -1, giving two sets of $n/2$ columns. Consider the columns as points and these two sets as blocks. It can be shown that we obtain a 3-(n, $n/2$, $n/4$ - 1) design, called a Hadamard 3-design. Any 3-design with the above parameters arises from a Hadamard matrix. Example 1.3 is the smallest Hadamard 3-design. Observe that any derived design of a Hadamard 3-design is a Hadamard 2-design. Conversely, if a Hadamard 3-design E is constructed as an extension of a Hadamard 2-design D, then D determines E. This is seen as follows. By Lemma 1.18 every derived design of E is a Hadamard 2-design with the same parameters as that of D. So any two blocks of E with non-empty intersection must intersect in (n/4 -1) + 1 = n/4 points. Since every block has at most one block disjoint from it, it is easily seen that every block has a unique block disjoint from it. The (unique) recipe of E from D is now obvious: For every block of D take its complement as a block and add an extra point ∞ to all the blocks of D.

Remarks 1.25. Let F be a field and V an (n + 1)-dimensional vector space over F. Then the set of all subspaces of V, ordered by inclusion, is called the n-dimensional projective space over F. If we take GF(q) to be the field F, then this is denoted by PG(n, q). The 1-dimensional, 2-dimensional, 3-dimensional and n-dimensional subspaces of V are respectively called the points, lines, planes and hyperplanes. In general the (i + 1)-dimensional subspaces are called the i-flats. We can form an incidence structure having respectively the points and i-flats of PG(n, q) as its points and blocks. This may be denoted by $PG_i(n, q)$ and it can be shown that $PG_i(n, q)$ forms a 2-design. Its parameters can be listed in terms of the Gaussian coefficient

$\begin{bmatrix} n \\ d \end{bmatrix}_q$ which denotes the number of d-dimensional subspaces of an n-dimensional vector space over GF(q). It is known that

$$\begin{bmatrix} n \\ d \end{bmatrix}_q = (q^n - 1)(q^{n-1} - 1)\cdots(q^{n-d+1} - 1) \,/\, (q^d - 1)(q^{d-1} - 1)\cdots(q - 1).$$

Then, the parameters of $PG_i(n, q)$ are

$$v = \begin{bmatrix} n+1 \\ 1 \end{bmatrix}_q = (q^{n+1} - 1)/ (q - 1), \quad b = \begin{bmatrix} n+1 \\ i+1 \end{bmatrix}_q, \quad r = \begin{bmatrix} n \\ i \end{bmatrix}_q,$$

$$k = \begin{bmatrix} i+1 \\ 1 \end{bmatrix}_q = (q^{i+1} - 1)/ (q - 1), \quad \lambda = \begin{bmatrix} n-1 \\ i-1 \end{bmatrix}_q.$$

The 2-design $PG_{n-1}(n, q)$ formed by the points and hyperplanes of PG(n, q) is a symmetric $((q^{n+1} -1)/(q-1), (q^n -1)/(q-1), (q^{n-1} -1)/(q-1))$ design.

Remarks 1.26. Closely related to the n-dimensional projective space PG(n, q) is the n-dimensional affine space AG(n, q). Here we take V to be an n-dimensional vector space over GF(q). Then the set of all cosets of subspaces of V ordered by inclusion forms AG(n, q). The cosets of {0} are called points, cosets of 1-dimensional subspaces are the lines, cosets of 2-dimensional subspaces are the planes and cosets of the (n-1)-dimensional subspaces are the hyperplanes of AG(n, q). In general the cosets of i-dimensional subspaces of V are called the i-flats of AG(n, q). We can form the incidence structure $AG_i(n, q)$ from the points and i-flats of AG(n, q). Then it can be shown that $AG_i(n, q)$ forms a 2-design with parameters

$$v = q^n, \quad b = q^{n-i}\begin{bmatrix} n \\ i \end{bmatrix}_q, \quad r = \begin{bmatrix} n \\ i \end{bmatrix}_q, \quad k = q^i, \quad \lambda = \begin{bmatrix} n-1 \\ i-1 \end{bmatrix}_q.$$

The design $AG_i(n, q)$ has the further property of being resolvable. That is, the blocks of $AG_i(n, q)$ can be partitioned into "parallel" classes, so that each point occurs exactly once in every parallel class. For example the $q^2 + q$ lines of the affine plane AG(2, q) can be partitioned into $q + 1$ parallel classes of q lines each so that each point occurs exactly once in each parallel class. This is also true in general: two hyperplanes of AG(n, q) are either disjoint or intersect in q^{n-2} points.

Remarks 1.27. A fundamental method of constructing 2-designs (and also the first systematic construction method) is due to Bose [20] and it is now known as the Method of differences. Here one starts with an abelian group (G, +). For a subset $B \subseteq G$, $|B| = k$, consider the k(k-1) ordered differences of its elements. We call a collection of m subsets $B_1, B_2, \ldots, B_m$ of G, a generating set of blocks or a set of initial blocks if among the mk(k-1) differences arising from these m blocks, every non-zero element of G occurs exactly a constant number λ times. If $\{B_1, B_2, \ldots, B_m\}$ is such a generating set of blocks, then taking the elements of G as the points and the family $\{B_j + g: g \in G\}$, $1 \leq j \leq m$ as blocks, we get a 2-design D with parameters $v = |G|$, k, and λ.

As an illustration of this method take $G = \mathbf{Z}_{13}$ under addition. Take $B_1 = \{1, 3, 9\}$ and $B_2 = \{2, 6, 5\}$. It can be checked that every non-zero element of G occurs exactly $\lambda = 1$ times amongst the differences from B_1 and B_2. Then we get a (13, 26, 6, 3, 1)-design.

Remark 1.28. There are other important methods based on (for instance) group theory, number theory, or coding theory which are used to construct designs. The reader is referred to Beth, Jungnickel and Lenz [15], Hall [74], Wallis [177], or Cameron and van Lint [49].

We give below a famous result of Cameron [43] concerning extendable symmetric 2-designs. We reproduce the proof given in Cameron and van Lint [49]. This is because we will often refer to this result and some ideas used in the proof later on.

Theorem 1.29. (Cameron) If a 3-(v, k, λ) design D is an extension of a symmetric 2-design, then one of the following holds:

(i) $v = 4(\lambda + 1)$, $k = 2(\lambda + 1)$;

(ii) $v = (\lambda + 1)(\lambda^2 + 5\lambda + 5)$, $k = (\lambda + 1)(\lambda + 2)$;

(iii) $v = 112$, $k = 12$, $\lambda = 1$;

(iv) $v = 496$, $k = 40$, $\lambda = 3$.

Proof. First observe that all the derived designs of D are symmetric designs by Lemma 1.18 and all of them have parameters v-1, k-1, λ. By Theorem 1.11, any two blocks of D are either disjoint or intersect each other in $\lambda + 1$ points (note that D has one point not in D_p). A 3-design D is then an extension of a symmetric 2-design if and only if any two blocks of D meet in 0 or $\lambda + 1$ points. A 3-design with these properties has $r = v-1$, $\lambda_2 = k - 1$, and $(v-2)\lambda = (k-1)(k-2)$.

Let B be a block of D. If $p, q \notin B$, then there are $k\lambda/(\lambda + 1)$ blocks containing p and q and meeting B in $\lambda + 1$ points, and hence $(k - \lambda - 1)/(\lambda + 1)$ blocks disjoint from B. This shows that the incidence structure D_0, whose points are those outside B and whose blocks are those disjoint from B, is a 2-$(v - k, k, (k - \lambda - 1)/(\lambda + 1))$ design. Applying Lemma 1.7 (the basic parameter relations of a 2-design), the number of blocks of D_0 is $(v - k)(v - k - 1)(k - \lambda - 1)/k(k - 1)(\lambda + 1)$.

If D_0 is degenerate with a single block, then $v = 2k$, and hence $v = 4(\lambda+1)$, $k = 2(\lambda + 1)$. Otherwise apply Fisher's inequality (Theorem 1.10) to D_0, giving $(v - k - 1)(k - \lambda - 1) \geq k(k - 1)(\lambda + 1)$, which simplifies to $(k - 1)(k - (\lambda + 1)(\lambda + 2)) \geq 0$ and hence $k \geq (\lambda + 1)(\lambda + 2)$. However, $b = v(v - 1)/k = (k^2 - 3k + 2\lambda + 2)(k^2 - 3k + \lambda + 2)/k\lambda^2$; so k divides $2(\lambda + 1)(\lambda + 2)$. If $k = 2(\lambda + 1)(\lambda + 2)$, then $\lambda = 1$ or 3, giving cases (iii) or (iv) of the theorem. If $k = (\lambda + 1)(\lambda + 2)$, then we have case (ii).

Exercise 1.30. Show that a 3-design is an extension of a symmetric design if and only if any two blocks of the design intersect in 0 or y points for some positive integer y.

Remarks 1.31. Designs occurring in case (i) of the above theorem are the so called Hadamard 3-designs, which exist infinitely often. Lam et al. [101] have ruled out (iii), which would be an extension of a projective plane of order 10. The design in (ii) for $\lambda = 1$ is the (unique) extension of the projective plane of order 4. Recently, Bagchi [8] has proved the non-existence of (ii) for $\lambda = 2$ using the ternary code of the design; the situation of $\lambda \geq 3$ is open. Nothing is known about case (iv).

II. STRONGLY REGULAR GRAPHS AND PARTIAL GEOMETRIES

Our aim in this chapter is to gather together some basic results from strongly regular graphs and partial geometries. These topics have had a profound influence in the area of combinatorial designs after Bose's classical paper [21] of 1963. The results of the first two chapters will provide the necessary background for later chapters. We refer to Harary [77] for the necessary background in graph theory. Marcus and Minc [109] will generally suffice for details of matrix results used. For further applications of matrix tools in a variety of problems on designs, we refer to M.S. Shrikhande [150].

Let Γ be a finite undirected graph on n vertices. The adjacency matrix A of Γ is a square matrix of size n. The diagonal entries of A are zero and for $i \neq j$, the (i, j) entry of A is 1 or 0 according as the vertices i and j are joined by an edge or not. There are other types of adjacency matrices used. For example in Seidel [143], or Goethals and Seidel [67], a $(0, \pm 1)$ adjacency matrix is used.

A graph Γ is called <u>regular</u> of valency a if A has constant row sum a. The adjacency matrix reflects many other graphical properties of Γ.

We now state two basic definitions. Firstly, a matrix A is <u>permutationally congruent</u> to a matrix B if there is a permutation matrix P such that $A = P^t BP$. In other words, A can be transformed to B by a simultaneous permutation of its rows and columns.

Secondly, an $n \times n$ matrix A $(n \geq 2)$ is called <u>reducible</u> (or decomposable) if it is permutationally congruent to a matrix of the form $\begin{bmatrix} B & C \\ 0 & D \end{bmatrix}$, where B and D are square matrices. The matrix A is <u>irreducible</u> (or indecomposable) otherwise.

The following two results from matrix theory are frequently used in algebraic graph theory.

Theorem 2.1. (Perron). Let A be an $n \times n$ non-negative indecomposable matrix. Then

(i) A has a real positive eigenvalue ρ which is a simple root of the characteristic polynomial of A.

(ii) If λ_i is any other eigenvalue of A, then $|\lambda_i| \leq \rho$.

(iii) There exists a positive eigenvector corresponding to ρ.

The eigenvalue ρ in the above theorem is called the maximal eigenvalue of A. Note also that a positive eigenvector means one with positive entries.

Theorem 2.2. (Frobenius). Let $A_1, A_2, \ldots, A_l$ be $n \times n$ real matrices. Then there exists an orthonormal basis consisting of common eigenvectors of $A_1, A_2, \ldots, A_l$ if and only if

(i) $A_1, A_2, \ldots, A_l$ are symmetric, and

(ii) $A_iA_j = A_jA_i$ for all $1 \leq i \neq j \leq l$.

Remark 2.3. Let A be the adjacency matrix of a regular graph having valency a. Then by Perron's theorem, a is an eigenvalue of A of multiplicity 1 provided Γ is connected. Any other eigenvalue λ satisfies $|\lambda| \leq a$. As an illustration of how useful these two theorems are, we give an elegant result of Hoffman [86]. This result is often used in algebraic graph theory.

Theorem 2.4. (Hoffman). Let Γ be a graph on n vertices with adjacency matrix A. Then there exists a polynomial $p(x)$ over the rationals such that $p(A) = J$, the all one matrix, if and only if Γ is regular and connected. In particular, if Γ is regular and connected and A has eigenvalue a (of multiplicity one), and other distinct eigenvalues $\beta_1, \beta_2, \ldots, \beta_t$ (with associated multiplicities) then,

$$p(x) = n \frac{\prod_{i=1}^{t} (x-\beta_i)}{\prod_{i=1}^{t} (a-\beta_i)}$$

Proof. Suppose $p(A) = J$. Then $AJ = Ap(A) = p(A)A = JA$. Thus, the $(i, j)^{th}$ entry of JA = valency of i = valency of j = $(i, j)^{th}$ entry of AJ. This means Γ is regular. Next suppose u and v are any two distinct vertices of A. Let $p(x) = \sum_{i=0}^{k} a_i x^i$, over the rationals. Then $p(A) = \sum_{i=0}^{k} a_i A^i = J$. This means that there is some i, $0 \leq i \leq k$, such that A^i has a non-zero entry in position (u, v). This then gives a path joining u and v, showing that Γ is connected. Next suppose Γ is regular of valency a and connected. Then A has a simple eigenvalue a and other distinct eigenvalues $\beta_1, \beta_2, \ldots, \beta_t$.
Now A, I, and J are pairwise commuting, symmetric matrices. Hence by Frobenius' Theorem 2.2, we can find an orthogonal matrix U such that

$$\mathbf{U}A\mathbf{U}^t = \text{diag}\,(a, \beta_1, \beta_1, \ldots, \beta_1, \beta_2, \ldots, \beta_2, \ldots, \beta_t, \beta_t, \ldots, \beta_t)$$
$$\mathbf{U}I\mathbf{U}^t = \text{diag}\,(1, 1, \ldots, 1) \text{ and}$$
$$\mathbf{U}J\mathbf{U}^t = \text{diag}\,(n, 0, 0, \ldots, 0).$$

Form the polynomial $p(x) = n \dfrac{\prod_{i=1}^{t} (x-\beta_i)}{\prod_{i=1}^{t} (a-\beta_i)}$.

We then verify that $\mathbf{U}p(A)\mathbf{U}^t = \mathbf{U}J\mathbf{U}^t$. This gives the desired conclusion $p(A) = J$.

The polynomial $p(x)$ in Theorem 2.4 is referred to as the Hoffman polynomial of the graph Γ.

We now define a strongly regular graph. This notion was formally introduced by Bose [21] in 1963. The concept is, however, implicit in the classical work on partially balanced incomplete block designs (PBIBDs) by Bose and Nair [26] . See also Bose and Shimamoto [27].

Definition 2.5. A graph Γ on v vertices is said to be strongly regular (SR) with parameters (v, a, c, d) if

(i) Γ is regular of valency a;

(ii) any two adjacent vertices are simultaneously adjacent to c other vertices;

(iii) any two non-adjacent vertices are simultaneously adjacent to d other vertices.

Remark 2.6. In the terminology of Bose and Nair [26], a strongly regular graph is exactly a two-class association scheme. In matrix language, a graph with (0, 1) adjacency matrix A is strongly regular with parameters (v, a, c, d) iff A satisfies the matrix equations $AJ = aJ$, $A^2 = aI + cA + d(J - I - A)$ for integral a, c, d.

The next lemma is easily proved by counting arguments.

Lemma 2.7. The complementary graph $\overline{\Gamma}$ of a strongly regular graph $\Gamma(v, a, c, d)$ is also strongly regular with parameters $(\overline{v}, \overline{a}, \overline{c}, \overline{d})$ given by $\overline{v} = v, \overline{a} = v - 1 - a, \overline{c} = v - 2a + d - 2, \overline{d} = v - 2a + c$.

We now give some examples of strongly regular graphs.

Example 2.8. The graphs , , , and the Petersen graph are all strongly regular

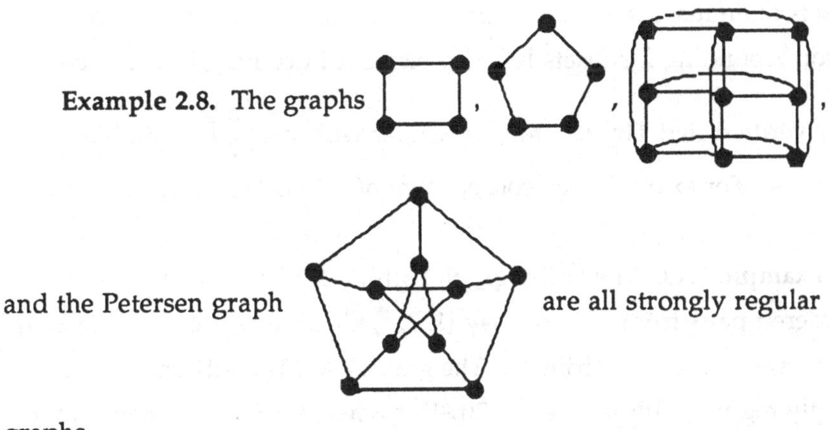

graphs.

Example 2.9. The complete graph K_n ($n \geq 3$) is trivially a strongly regular graph with v = n, a = n-1, c = n-2, d = 0. More generally Γ(m, n), the disjoint union of m complete graphs K_n, is strongly regular with parameters v = mn, a = n-1, c = n-2, d = 0. It is easily shown that any strongly regular graph with d = 0 is of this type.

Example 2.10. The complementary graph of Γ(m, n) is known as the complete m-partite graph and is usually denoted by K(m, n). In K(m, n) the vertex set is partitioned into m classes of n vertices each. Vertices in the same class are non-adjacent and any two vertices from different classes are adjacent. By Lemma 2.7, the graph K(m, n) is strongly regular.

Exercise 2.11. Show that a = d holds iff Γ = K(2, a).

Example 2.12. The triangular graph $T(n)$, $n \geq 3$, has as vertices all 2 -subsets of an n-set, say $\{1, 2, ..., n\}$. Two vertices of $T(n)$ are adjacent iff the corresponding 2-subsets have an element in common. It can be easily verified that $T(n)$ is strongly regular with $v = \binom{n}{2}$, $a = 2(n-2)$, $c = n-2$, $d = 4$. For example, the complement of $T(5)$ is the Petersen graph.

Example 2.13. The lattice graph $L_2(n)$, $n \geq 2$ has as its vertices the n^2 ordered pairs from an n-set, say $\{1, 2,..., n\}$. Two vertices are adjacent iff they agree in one coordinate. The graph $L_2(n)$ is easily seen to be strongly regular with $v = n^2$, $a = 2(n-1)$, $c = n-2$, $d = 2$. It is obvious that the third graph in Example 2.8 is the graph $L_2(3)$.

Example 2.14. The Clebsch graph has as its vertex set all the subsets of $\{1, 2, 3, 4, 5\}$ having even cardinality. Two vertices are adjacent iff the symmetric difference of the corresponding subsets has size 4. The Clebsch graph is strongly regular with $v = 16$, $a = 5$, $c = 0$, $d = 2$.

Before giving another rich source for examples of strongly regular graphs we need some background from permutation groups. Let G be a permutation group acting on a finite set of X. For $x \in X$, the orbit xG of x is the set of distinct elements of the form xg, $g \in G$. The stabilizer of x is the set of all $g \in G$ which fix x. The stabilizer of x is a subgroup of G, and is usually denoted by G_x. The fundamental relation : $|G| = |G_x|\ |xG|$ can be easily proved by use of Lemma 1.4. The number t of orbits of G on X is given by a result of Burnside: $t|G| = \sum |F(g)|$, where $F(g) = \{x \in G: xg = x\}$ is the set of fixed points of $g \in G$. The group G is called a transitive permutation group if there is precisely one orbit under the action of G on X. Thus for a transitive group G the above results reduce to: $|G| = |G_x|\ |X|$ and $|G| = \Sigma|F(g)|$.

The rank r of a transitive group G is the number of orbits of the stabilizer G_x on X. Another result of Burnside gives a formula to find r: $r|G| = \Sigma|F(g)|^2$. These results can all be found in any standard reference on permutation groups, e.g., Wielandt [178] or Biggs[16]. Now let G be a transitive permutation group acting on X. We can let G act on X x X by defining $(x, y)g = ((x)g, (y)g)$ and suppose $D_0, D_1, ..., D_{s-1}$ denote the orbits under this action. Fix any $x \in X$, and set $D_i(x) = \{y \in X: (x, y) \in D_i\}$. Thus $D_i(x)$, $i = 0, 1, 2, ..., s-1$ are the orbits of G_x on X. So the rank r of G equals s and we may take $D_0(x) = \{x\}$. It is also easy to see that if D is any orbit of G on X x X, then $D^t = \{(x, y) \in X \times X: (y, x) \in D\}$ is also an orbit of G on X x X. The connection between graphs and permutation groups is then given by the following lemma, which is well known.

Lemma 2.15. Suppose G is transitive and an orbit $D \neq D_0$ of G on X x X, satisfies $D^t = D$. Then we can form a graph (X, E) by defining adjacency between distinct $x, y \in X$, whenever $(x, y) \in D$. The group G acts as a vertex-transitive and edge-transitive group of automorphisms on the graph (X, E). In particular if G is transitive and has even order, there exists an undirected graph (X, E) on which G acts as a vertex-transitive group of automorphisms.

Proof. The first part is obvious. Suppose further that G is transitive of even order. Then G contains an involution which interchanges a pair of elements $x, y \in X$. If D is an orbit containing (x, y), then D contains (y, x). Thus $D^t = D$ and the result follows.

To illustrate the above ideas, we give the following well known example (see Biggs [16]).

Example 2.16. Let the symmetric group S_n act on the set X of all 2-subsets of $\{1, 2, ..., n\}$, $n \geq 4$. Then clearly S_n acts transitively on X. Writing ij for $\{i, j\}$, S_n has the following orbits on X x X: The diagonal D_0 consisting of all pairs (ij, ij), D_1 consisting of all (ij, kl), where i, j, k, l are distinct, and D_2 consisting of all (ij, il), when i, j, l are distinct. Here $D_1 = D_1{}^t$ and $D_2 = D_2{}^t$, and thus we obtain two complementary graphs on which S_n acts. The graph (X, E) in the case n = 5 is the Petersen graph:

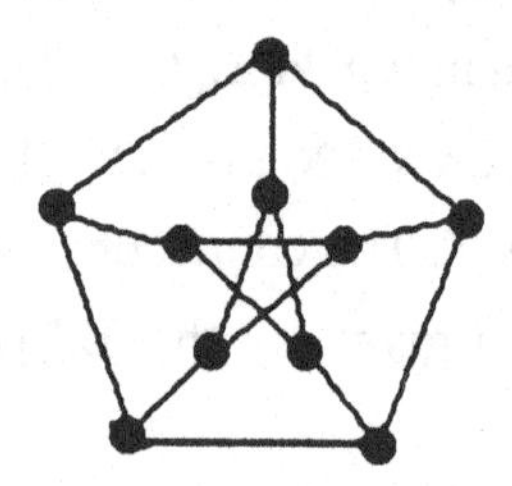

Exercise 2.17. Show that the S_n action (on D_2) described as above produces the triangular graphs of Example 2.12.

The above example motivates the following definition.

Definition 2.18. Let G be a transitive permutation group of even order, acting on a finite set X. Then G is called a rank 3 group if G has exactly 3 orbits: the diagonal D, and (say) O_1, O_2 on X x X (equivalently G_x has 3 orbits on X). From such a rank 3 group we can form a graph on X by defining an edge between x, y whenever $(x, y) \in O_1$ (say). Such a graph is called a rank 3 graph. Rank 3 graphs are important to us because of the next well known lemma.

Lemma 2.19. A rank 3 graph is strongly regular.

Proof. Let G be the rank 3 group which produces a rank 3 graph

(X, E). This graph is regular since G is transitive. Further since G is transitive on adjacent, as well as on non-adjacent, pairs of vertices, the graph (X, E) is strongly regular. The parameters of this graph can be easily found.

Remark 2.20. As already mentioned the graphs T(n) and $L_2(n)$ are rank 3 graphs. There are examples of strongly regular graphs which are not rank 3 graphs. One such example is given in Exercise 2.30.

As an example of rank 3 graphs, we give the well known construction of the Paley graphs P(q) (see Tonchev [175]).

Lemma 2.21. Let X = GF(q), where q is a prime power, $q \equiv 1 \pmod 4$. Take the elements of X as vertices: Form a graph P(q) by defining adjacency between a pair of distinct elements of X iff their difference is a square in GF(q). Then P(q) is a rank 3 graph and hence strongly regular with parameters $v = q$, $a = (q - 1)/2$, $c = (q - 5)/4$, $d = (q - 1)/4$.

Proof. Let G be the permutation group acting on GF(q) consisting of all linear transformations of the form $y = \alpha^2 x + \beta$, $\alpha, \beta \in GF(q)$, $\alpha \neq 0$. Since for any GF(q), the group of affine transformations $y = a x + b$, $a, b \in GF(q)$, $a \neq 0$ is 2-transitive, it can be shown from elementary permutation group theory (see, e.g. [16]) that $|G| = q(q - 1)/2$. Since G contains all translations $y = x + \beta$, $\beta \in GF(q)$, the group is transitive. The stabilizer G_0 of zero consists of all transformations of the type $y = \alpha^2 x$, $\alpha \neq 0$. Thus G_0 divides X into 3 orbits: {0}, the set of $(q-1)/2$ squares of GF(q), and the set of $(q-1)/2$ non-squares of GF(q). Thus G is a rank 3 group and hence the above construction of P(q) gives a strongly regular graph. The graph P(q) is obviously strongly regular of valence $a = (q-1)/2$. To find the other parameters of P(q) we can proceed as follows. Define a matrix $Q = (q_{ij})$ indexed by the elements of GF(q). Define $q_{ii} = 0$ and for $i \neq j$, $q_{ij} = 1$ or -1 according as j - i is a square in GF(q) or not. It can be checked that $QQ^t = qI - J$ and $QJ = JA = O$. Since $q \equiv 1 \pmod 4$, the matrix Q is symmetric. It then follows that $A = (1/2)(Q + J - I)$ is the adjacency matrix of P(q). The matrix equations $AJ = [(q-1)/2]J$ and $A^2 = A + [(q-1)/4]I + [(q-1)/4]J$ then give the parameters of P(q) to be $v = q$, $a = (q-1)/2$, $c = (q-5)/4$ and $d = (q-1)/4$.

There are other constructions of strongly regular graphs from groups.

Hoffman's result, Theorem 2.3 was used by S. S. Shrikhande and Bhagwandas [162] to prove the following spectral characterization of strongly regular graphs. Since we shall often use this result in the sequel, we give a proof. See also Goethals and Seidel [67]. A null graph is one for which the adjacency matrix A is the zero matrix. A complete graph is one for which $A = J - I$.

Theorem 2.22. A regular, connected non-complete adjacency matrix A of size v having valence a is strongly regular with parameters (v, a, c, d) if and only if A has exactly 3 distinct eigenvalues $\theta_0 = a$, θ_1 and θ_2. Moreover then $c = a + \theta_1 + \theta_2 + \theta_1\theta_2$ and $d = a + \theta_1\theta_2$.

Proof. Let $A \neq 0$, and let J-I be strongly regular with parameters (v, a, c, d). Equivalently, $A^2 = aI+cA+d(J-I-A)$. Let $\theta \neq a$ be an eigenvalue of A. By Frobenius' Theorem 2.2, θ satisfies the quadratic equation $\theta^2+(d-c)\theta+d-a = 0$. The discriminant is > 0, unless c=d=a. However, c=d=a is impossible since always $c \leq a-1$. Hence A has 3 distinct eigenvalues.

Conversely, suppose $A \neq 0$, $A \neq J-I$ is regular, connected, and has 3 distinct eignenvalues $\theta_0 = a$, θ_1, and θ_2. By Hoffman's result, p(A)=J, where

$$p(x) = \frac{v\,(x-\theta_1)(x-\theta_2)}{(a-\theta_1)(a-\theta_2)}.$$

Thus $(A-\theta_1)(A-\theta_2) = [(a-\theta_1)(a-\theta_2)/v]J$.

Simplifying this we get the matrix equation (*)

(*) $$A^2 = \theta_1\theta_2 I + (\theta_1+\theta_2)A + [(a-\theta_1)(a-\theta_2)/v]J.$$

Therefore,

$$A^2 = aI + cA + d(J-I-A),$$

on putting $a-d = \theta_2$, $c-d = \theta_1+\theta_2$, $d=[(a-\theta_1)(a-\theta_2)/v]$.

In (*) equate diagonal entries which are 0, off diagonal entries which are 1, and then off-diagonal entries which are 0. It then follows that c and d are integers. Thus A is a strongly regular graph with parameters (v, a, c, d) expressible in terms of the eigenvalues of A. This completes the proof.

The following result gives the well known "rationality conditions" for strongly regular graphs. These are necessary conditions which the parameters must satisfy.

Theorem 2.23. Let Γ be a strongly regular graph with parameters (v, a, c, d). Then

(i) $v = 4d+1$, $a = 2d$, $c = d-1$, or

(ii) $(c-d)^2+4(a-d)$ is a square and the expressions $m_i =$

$$\left(\frac{1}{2}\right)\left[(v-1) \pm \frac{(v-1)(d-c)-2a}{\sqrt{(c-d)^2+4(c-d)}}\right]$$ (i=1, 2) are non-negative integers.

Proof. If $d = 0$ then Γ is a complete bipartite graph whose parameters satisfy (ii). Hence assume $d \neq 0$. Let $\theta \neq a$ be an eigenvalue of the adjacency matrix A. Then, as in the proof of the previous theorem, $\theta^2+(d-c)\theta +d-a = 0$, and the eigenvalues θ_1, θ_2 different from a are given by $\theta_1, \theta_2 = ((c-d)\pm\sqrt{\Delta})/2$ where $\Delta= (c-d)^2+4(a-d) \neq 0$. Let m_1, m_2 be the multiplicities of θ_1, θ_2. Then $m_1+m_2 = v-1$ and $a+m_1\theta_1+m_2\theta_2 = 0$. Using the values of θ_1 and θ_2 we obtain $(m_1-m_2)\sqrt{\Delta} = (m_1+m_2)(d-c) - 2a$. So if Δ is not a square, then $m_1 = m_2 = (v-1)/2$, giving $2a = (v-1)(d-c)$. Now $d-c \geq 1$ and if $d-c \geq 2$ then $v-1 \leq a$ which is impossible. Thus $d-c = 1$ giving $a = 2d$ which is case (i) of the conclusion of Theorem 2.18. If Δ is a square, then solving for m_1 and m_2 leads to the stated values in part (ii). This completes the proof.

Exercise 2.24. Give a matrix proof of Lemma 2.7.

Remark 2.25. The graphs T(n) and $L_2(n)$ are of type (ii), the graph $L_2(3)$ is also of type (iii). The Paley graphs P(q) of Lemma 2.21 are of type (i).

There are some famous "characterizations by parameters" results dealing with strongly regular graphs. We mention them below.

Theorem 2.26. Let Γ be a strongly regular graph with $v = \binom{n}{2}$, $a = 2(n-2)$, $c = n-1$, and $d= 4$ $(n \geq 3)$. If $n \neq 8$, then Γ is isomorphic to the triangular graph T(n). If $n = 8$, then beside T(8), there are exactly 3 non-isomorphic strongly regular graphs with the same parameters as T(8).

Remark 2.27. Connor [56] proved the above result for $n>8$; Hoffman [88] and Chang [54] settled $n = 7$; S.S. Shrikhande [159] proved the cases n = 5, 6; Hoffman [88] and Chang [54] proved that there are 3 exceptional graphs.

A result proved by S.S. Shrikhande [160] with the same flavor concerns $L_2(n)$.

Theorem 2.28. Let Γ be a strongly regular graph with parameters $v = n^2$, $a = 2(n-1)$, $c = n-2$, and $d= 2$ $(n \geq 2)$. If $n \neq 4$, then Γ is isomorphic to $L_2(n)$. If n = 4, then there are exactly two non-isomorphic graphs with the same parameters as $L_2(4)$.

Remark 2.29. The exceptional graph in the above theorem is

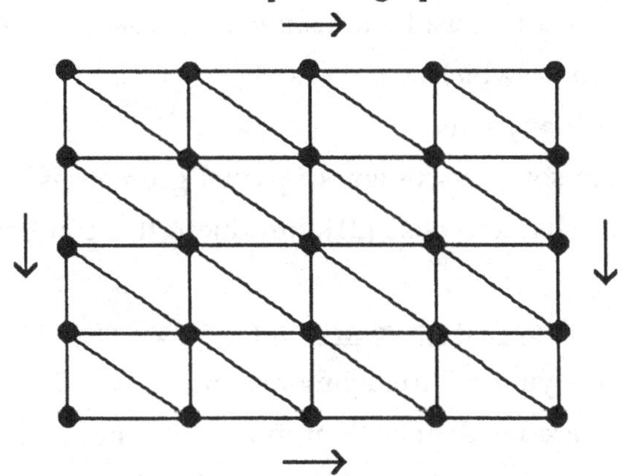

Here the vertices and edges are identified with those on the opposite sides of the figure as shown by the arrows. Equivalently it can be thought of as drawn on a torus. For example, observe that every vertex has 6 vertices adjacent to it: two each on vertical and horizontal lines and two on the diagonal as drawn in the figure. Also the four points on the corners actually correspond to a single vertex.

Exercise 2.30. Proceed through the following steps to show that the exceptional graph Γ with the same parameters as that of $L_2(4)$ is <u>not rank 3.</u>

(a) The neighborhood of any vertex x of Γ is a hexagon with x as a center and forming 6 triangles. Also of the remaining 9 vertices, three are adjacent to opposite pairs of vertices of this hexagon and the remaining 6 are adjacent to edges of the hexagon.

(b) Let G be the full automorphism group of Γ. Then G is transitive on Γ. Further, if $H = G_x$, then H is faithful on the neighborhood of x. Also, H contains a 6-cycle and a permutation of order two fixing two opposite vertices (and reflecting the other vertices of the hexagon). Conclude that H is isomorphic to D_6, the dihedral group on 6 letters.

(c) If G was rank 3, then H must be transitive on the set of 9 vertices not in the neighborhood, which is impossible.

(d) Show that G is actually rank 4.

We now come to some discussion of partial geometries which is the third topic in this chapter. Bose [21] introduced this concept in 1963.

Definition 2.31. A <u>partial geometry</u> (r, k, t) is a system D of points and lines (blocks) satisfying the following axioms:

(1) Any two points are incident with at most one common line.

(2) D is a 1-design, i.e., every line of D has k points and every point is contained in r lines.

(3) If a point x is not incident with a line X, there exist exactly t lines through x incident with X ($1 \le t \le r$, $1 \le t \le k$).

The following are examples of partial geometries.

Examples 2.31.

(i) The points and lines of an affine plane AG(2, q) form a partial geometry with $r = q+1$, $k = q$, $t = q$.

(ii) The points and lines of a projective plane PG(2, q) is a partial geometry with $r = k = t = q+1$.

(iii) Any design with parameters v, b, r, k, $\lambda = 1$ is a partial geometry r, k, $t = k$ by taking blocks as the lines.

(iv) The lattice graph $L_2(n)$ is a partial geometry with n^2 points and the 2n rows and columns of the array are the lines. Here $r=2$, $k = n-1$, $t=1$.

(v) The triangular graph T(n) forms a partial geometry with $r=2$, $k = n-1$, $t=2$.

(vi) A large class of partial geometries is given by nets defined later in this chapter.

From a partial geometry (r, k, t) we can form a graph G in the following way. Take the vertices of G to be the points of the geometry. Join two vertices of G by an edge if the corresponding points lie on a line of the geometry. The graph of a partial geometry (r, k, t) is called a geometric graph (r, k, t).

In [21], Bose proved the following theorem by elementary counting arguments.

Theorem 2.32. The graph G of a partial geometry (r, k, t) is strongly regular with parameters $v = k[(r-1)(k-1) + t]/t$, $a=r(k-1)$, $c = [(t-1)(r-1) + k-2]$, and $d = rt$. Bose [21] calls a strongly regular graph G pseudo geometric if it has the parameters of a geometric graph (r, k, t). The exceptional cases arising in Theorem 2.23 dealing with the triangular graph T(8) or Theorem 2.25 for the lattice graph $L_2(4)$ provide examples of pseudo-geometric graphs which are not geometric. In fact the main aim of Bose's paper was to obtain a unifying theorem which as a special case gives the characterization theorems for T(n), L_2 (n), and nets defined below. Using elegant "claws and cliques" methods, Bose [21] proved the following fundamental result:

Theorem 2.33. A pseudo-geometric (r, k, t) graph G is a geometric graph (r, k, t) if $k \geq f(r, t) = \frac{1}{2}[r(r-1) + t(r+1)(r^2-2r+2)]$.

Bose introduced the concept of a partial geometry (r, k, t) as a generalization of Bruck's notion of a net [36] . We close this chapter with a brief introduction to nets.

Definition 2.34. A <u>net</u> N of degree r is a system of points and lines (blocks) satisfying the following: The lines of N can be partitioned into r mutually disjoint non-empty parallel classes such that two lines of different parallel classes intersect in one point and every point of N is incident with precisely one line of each parallel class.

Exercise 2.35. Show that, if N is a net with $r \geq 3$ parallel classes then N is a 1-design with block size (some) k and replication number r. Also show that in that case every parallel class has k lines and N has k^2 points in all.

Exercise 2.36. Show that, if N is a net with $r \geq 3$ parallel classes, then N is a partial geometry (r, k, r-1). Also show that a net with three parallel classes can always be constructed for any $k \geq 2$.

The most trivial examples of nets are affine planes AG(2, q) where $r = q + 1$ and $k = q$. In fact, a net can be thought of as an affine plane with some parallel classes removed, though there exist examples of nets that cannot be obtained in this way (See [15]). In this connection, we note that some of the embedding theorems of Bruck [36] are not completely apparent in Bose's treatment [21] and we refer the reader to the book of Denes and Keedwell [62] for a lucid account of Bruck's results and related topics.

Exercise 2.37. Show that a net is a 2-design iff it is an affine plane.

Concluding Remarks 2.38.

(i) A partial geometry with $t = 1$ is called a generalized quadrangle. Here one has a large number of quadrangles but no triangles. This area, which strongly depends on constructions involving groups, is the topic of Payne and Thas [121].

(ii) Observe that all the examples of partial geometries given so far have the property that $t = 1$, k, or r-1. Such partial geometries are called improper. It was only in 1973 that proper partial geometries were discovered by Thas [169, 170] and we refer the reader to Batten [12] for a discussion. After the publication of the above mentioned papers of Thas, some new partial geometries have also been found and we refer to Haemers [72] for details.

(iii) We wish to impress upon the reader that the topic of strongly regular graphs is an important area in design theory and particularly so in the later chapters of this monograph. We mention three survey articles dealing with different aspects of strongly regular graphs: Hubaut [91], Seidel [144], and Brouwer and van Lint [33].

III. BASIC RESULTS ON QUASI-SYMMETRIC DESIGNS

Suppose D is a t-(v, k, λ) design with blocks $B_1, B_2, \ldots, B_b$. The cardinality $|B_i \cap B_j|$, $i \neq j$, is called an intersection number of D. Assume that $x_1, x_2, \ldots, x_s$ are the distinct intersection numbers of the design D. Specifying some of the x_i's or the number s can sometimes provide very useful information about the design. For instance, any 2-design with exactly one intersection number must necessarily be symmetric. Any 2-design with exactly the two intersection numbers 0 and 1 must be a non-symmetric 2-(v, k, 1) design.

In this chapter, we discuss designs which are in a sense "close" to symmetric designs. These are t-(v, k, λ) designs with exactly two intersection numbers. Such designs are called quasi-symmetric. We believe this concept goes back to S.S. Shrikhande [158] who considered duals of designs with $\lambda = 1$. We let x, y stand for the intersection numbers of a quasi-symmetric design with the standard convention that $x < y$.

Before proceeding further, we list below some well known examples of quasi-symmetric designs.

Example 3.1. Let D be a multiple of a symmetric 2-(v, k, λ) design. Then D is a quasi-symmetric 2-design with $x = \lambda$ and $y = k$.

Example 3.2. Let D be a 2-(v, k, 1) design with $b > v$. Then obviously D is quasi-symmetric with $x = 0$ and $y = 1$.

Example 3.3. Let D be a quasi-residual 2-design with $\lambda = 2$. That D is quasi-symmetric is well known [74] and can be shown easily as follows: The parameters of such a D are $v = \binom{k+1}{2}$, $b = \binom{k+2}{2}$,

$r = k + 2, k, \lambda = 2$. Fix a block B of D, and let n_i be the number of blocks $C \neq B$ which meet B in exactly i points ($i = 0, 1, ..., k$). Then $\Sigma\, n_i = b-1$, $\Sigma\, in_i = k(k + 1)$, and $\Sigma i(i - 1)n_i = k(k - 1)$ follow by two-way counting. We then obtain $\Sigma\, (i - 1)(i - 2)n_i = 0$, which implies $n_i = 0$, for $i \neq 1, 2$. Thus D is quasi-symmetric with $x = 1$ and $y = 2$.

The following examples need some advanced knowledge and may be skipped at the first reading. We shall discuss these topics further in Chapter VI.

Example 3.4. The well-known Witt 3-(22, 6, 1) design D is the extension of PG(2, 4). Thus D is quasi-symmetric with intersection numbers 0 and 2.

Example 3.5. The Witt design with parameters 4-(23, 7, 1), which is the two fold extension of PG(2, 4), is quasi-symmetric with $x = 1$ and $y = 3$. Its residual design is a 3-(22, 7, 4) design which is quasi-symmetric with intersection numbers $x = 1$, $y = 3$.

Example 3.6. A design D is called strongly resolvable if its blocks can be partitioned into c classes $\mathbf{B}_1, \mathbf{B}_2, ..., \mathbf{B}_c$ such that (i) the restriction of D to any $\mathbf{B}_i$ is a 1-design, i.e., every point occurs the same number of times on blocks of every class, and (ii) there are constants α and β such that two blocks intersect each other in α (respectively β) points if and only if they are in the same class (different classes).

In fact, a design satisfying (i) also satisfies (ii) iff $b = v - c + 1$, as was shown by S.S. Shrikhande and Raghavarao [163] and Hughes and Piper [93]. Clearly, such designs are quasi-symmetric. Typical examples are affine designs and Hadamard 3-designs introduced in Chapter I.

We now give a result of Bose [21] which concerns the dual partial geometry obtained from a BIBD with $\lambda=1$ and $b > v$.

Proposition 3.7. Let D be a 2-(v, k, 1) design with $b > v$. Let Γ be the geometric graph obtained from D by taking the blocks as vertices and joining two vertices by an edge if the corresponding blocks intersect. Then Γ is a strongly regular graph on b vertices and has $a = k(r-1)$, $c = (r-2)+(k-1)^2$, $d = k^2$.

The above result is immediate from Bose's Theorem 2.32 in Chapter II. Note that Γ is non-null since $r > 1$ and non-complete since $b > v$. This construction works if D is replaced by any quasi-symmetric design. This is a result of S.S. Shrikhande and Bhagwandas [162]. See also Goethals and Seidel [67].

Theorem 3.8. Let D be any quasi-symmetric 2-(v, k, λ) design with intersection numbers x and y. Form the <u>block graph</u> Γ by taking as vertices the blocks of D. Join two vertices of Γ by an edge whenever the corresponding blocks intersect in y points. Assume that Γ is connected. Then Γ is a strongly regular graph (n, a, c, d). The parameters of Γ are given by $n = b$, $a = \frac{k(r-1)-x(b-1)}{y-x}$, $c = a + \theta_1 + \theta_2 + \theta_1\theta_2$, $d = a + \theta_1\theta_2$, where $\theta_1 = (r - \lambda - k + x)/(y - x)$ and $\theta_2 = -\frac{k-x}{y-x}$.

Proof. Let N and A denote respectively the usual vxb incidence matrix of D and the adjacency matrix of Γ. Then $N^tN = kI + yA + x(J - J - A)$. Now NN^t has a simple eigenvalue rk and the remaining v - 1 eigenvalues $r - \lambda$ with multiplicity $v - 1$. Hence N^tN has eigenvalues rk with multiplicity 1, $r - \lambda$ with multiplicity $v - 1$, and 0 with multiplicity $b - v + 1$. Thus A has only 3 distinct eigenvalues θ_0,

θ_1, and θ_2, which using Theorem 2.21 makes Γ a strongly regular graph (n, a, c, d). The values of θ_0, θ_1, θ_2 and hence a, c, d can be easily shown to be the above stated values. This completes the proof of the theorem.

The following well known little observation is very useful in studying quasi-symmetric designs [150].

Corollary 3.9. In any quasi-symmetric 2-(v, k, λ) design D, $y - x$ divides both the integers $k - x$ and $r - \lambda$.

Proof. From the above proof the eigenvalues θ_1 and θ_2 are roots of the monic polynomial $\theta^2 + (d - c)\theta + d - a = 0$, which has integer coefficients. This implies that the rational numbers θ_1, θ_2 are in fact integers, thereby proving the required assertion.

Exercise 3.10. Show that if G is a strongly regular graph with parameters n, a, c and d, then the following are equivalent:

(i) G is a complete bipartite graph K(2, a).

(ii) a=d.

(iii) One eigenvalue of G is 0.

Also show that if these conditions do not hold, then one eigenvalue of G is positive and the other is negative.

Hint: The eigenvalues satisfy $\theta^2 - (c - d)\theta - (a - d) = 0$. Now use Exercise 2.11.

Exercise 3.11. Consider the complement $\overline{\Gamma}$ of the block graph Γ of a quasi-symmetric design D. Show that the parameters of $\overline{\Gamma}$ are

$$\overline{a} = \frac{y(b-1) - k(r-1)}{y-x}, \quad \overline{\theta}_1 = -\frac{r-\lambda-k+y}{y-x}, \quad \overline{\theta}_2 = \frac{k-y}{y-x}.$$

Hint: In Theorem 3.8, x and y can be interchanged.

Exercise 3.12. Let D be a quasi-symmetric design and let $\Gamma, \overline{\Gamma}$ have the same meaning as before. Then Γ (or $\overline{\Gamma}$) never satisfies the conditions of Exercise 3.10. In particular, θ_1 and $\overline{\theta}_1$ are never zero.

Hint: Consider Γ. If $\theta_2 = 0$, then $y \geq k = x$ implies $k = y = x$, which is impossible since D is proper. If $\theta_1 = 0$, then Exercise 3.10 shows that $c = a - 1$, which by Theorem 3.8 implies $\theta_2 = 1$, i.e., $k = y$. Since $x \neq y$, we must have $\Gamma = K(2, 1)$ which is impossible. A similar proof works for $\overline{\Gamma}$.

While no matrix-free proof of Corollary 3.9 is known to the authors, the following weaker version of Theorem 3.8 can be proved:

Exercise 3.13. Let X and Y be two blocks intersecting in y points. Write a_{ij} to denote the number of blocks intersecting X in i points and Y in j points (where i, j = x or y). Then we can obtain four linear equations in four unknowns, where the L.H.S. of these equations are $a_{xx} + a_{xy} + a_{yx} + a_{yy}$, $xa_{xx} + xa_{xy} + ya_{yx} + ya_{yy}$, $xa_{xx} + ya_{xy} + xa_{yx} + ya_{yy}$, and $x^2a_{xx} + xya_{xy} + yxa_{yx} + y^2a_{yy}$, where the R.H.S. of the last equation is $(k - y)^2\lambda + 2y(k - y)(\lambda - 1) + y(r - 2) + y(y - 1)(\lambda - 2)$ (justify!).

It is easily seen that these equations are linearly independent. Replacing y by x in the above finishes the proof.

Remarks 3.14. The fact that θ_1 and θ_2 of Theorem 3.8 always have integral multiplicities follows easily. Thus the "rationality conditions" in Theorem 2.22 give no additional restrictions on the parameters of the quasi-symmetric design. Another easily proven result is the following: If D is a quasi-symmetric 2-(v, k, λ) design, then so is the complementary design $\overline{D}$. Furthermore, the block graphs of D and $\overline{D}$ are isomorphic.

The following important and very useful property of quasi-symmetric designs is due to Cameron and van Lint [49, p. 27 - 28].

Theorem 3.15. In a quasi-symmetric design with connected block graph, and without repeated blocks, $b \leq \binom{v}{2}$.

Proof. Define a $\binom{v}{2}$ xb matrix M with entries 0 and 1 by $M = [m_{ij}]$ where $m_{ij} = 1$ if the i-th point-pair is in the j-th block and $m_{ij} = 0$ otherwise. Our proof is based on the fact that M^tM is non-singular. In that case, the rank of M is b which can not exceed $\binom{v}{2}$, the number of rows of M. We have $M^tM = \binom{k}{2}I + \binom{y}{2}A + \binom{x}{2}(J - I - A)$ where A is the adjacency matrix of the block graph. Clearly, then a zero eigenvalue of M^tM must correspond to an eigenvalue θ of A, where θ equals $-\dfrac{k(k-1) - x(x-1)}{y(y-1) - x(x-1)}$. Since k, x, y are all distinct by assumption, this eigenvalue is negative and must equal (by Exercise 3.12 and Theorem 3.8) $-\frac{k-x}{y-x} = \theta_2$. Hence $k + x - 1 = y + x - 1$, i.e., $k = y$ which is a contradiction.

We also give Cameron and van Lint's [49] proof of Theorem 3.15.

Let $M(\epsilon_0, \epsilon_1, \epsilon_2)$ be a $\binom{v}{2} \times b$ matrix with rows and columns corresponding to point-pairs and blocks respectively. Write $M = [m_{ij}]$ where m_{ij} equals ϵ_p if j-th block contains p points from i-th point-pair, p = 0, 1, 2. A careful counting produces the following equation:

$$M^t(0, 0, 1)M(\epsilon_0, \epsilon_1, 0)$$
$$= (y(k-y)\epsilon_1 + \binom{k-y}{2}\epsilon_0)A + (x(k-x)\epsilon_1 + \binom{k-x}{2}\epsilon_0)(J - I - A).$$

Clearly, if $b > \binom{v}{2}$, then the rank of both the matrices on the L.H.S. can not exceed $\binom{v}{2}$ and hence the L.H.S. is singular for any value of ϵ_0 and ϵ_1. Substitution of $(\epsilon_0, \epsilon_1) = (0, 1)$ and $(1, 0)$ gives two equations which show that the assumption leads to a contradiction.

In fact, Cameron and van Lint [49] claim a strengthening of Theorem 3.15.

Theorem 3.16. Let D be a quasi-symmetric design without repeated blocks and with $4 \le k \le v-4$. Then $b \le \binom{v}{2}$ with equality if and only if D is a 4-design.

The next two results are well known (See e.g. Cameron and van Lint [49]). They provide some useful parameter relationships and inequalities of quasi-symmetric designs, assuming particular values of the intersection numbers.

Proposition 3.17. Let x, y be the intersection numbers of a quasi-symmetric design with parameters v, b, r, k, λ.

(i) If $x = 0$, then $(r - 1)(y - 1) = (k - 1)(\lambda - 1)$.

(ii) If $x = 0$, then $b \le v(v - 1)/k$ and $y \le \lambda$.

(iii) If $x = 0$, $y = 1$ then $b = v(v - 1)/k(k - 1)$.

Proof. (iii) follows since $(x, y) = (0, 1)$ if and only if $\lambda = 1$. The usual parameter relations $bk = vr$, $\lambda(v - 1) = r(k - 1)$ then prove (iii). Next, consider (i) and assume that the quasi-symmetric design D has $x = 0$. Choose any point p and consider the "derived configuration" D_p.

Here, the points of D_p are those of D other than p and the blocks of D_p are the blocks of D through p. Then D_p has v - 1 points, r blocks, block size k - 1 and there are λ blocks of D_p through any point of D_p. Since $x = 0$, then the dual of D_p is a design with the usual parameters (r, v - 1, k - 1, λ, y - 1). Then the relation $(r - 1)(y - 1) = (k - 1)(\lambda - 1)$ is immediate. Applying Fisher's inequality to the dual of D_p gives $v - 1 \geq r$. This in turn gives $b \leq v (v - 1)/k$. That $y \leq \lambda$ follows from $(r - 1) (y - 1) = (k - 1)$ $(\lambda - 1)$ and $r \geq k$.

Corollary 3.18. Let D be a 2-(v, k, λ) design with $2 < k < v - 1$. Then any two of the following three statements imply the third and imply also that D is an extension of a symmetric 2- design (Consequently the conclusions of Cameron's Theorem 1.29 hold).

(i) D is a 3-design,
(ii) D is a quasi-symmetric design with $x = 0$,
(iii) $b = v(v - 1)/k$.

Proof. We observe that these three statements imply respectively the following three statements about D_p:

(i) D_p is a 2-design;
(ii) the dual of D_p is a 2-design;
(iii) D_p has equally many points and blocks.

In all the situations D_p is a 1-design. Then the various characterizations show that, if any two of (i), (ii), and (iii) hold, then D_p must be a symmetric design. Hence D is an extension of a symmetric design. Use of Theorem 1.29 then completes the proof.

Before proceeding further, we recall that a ladder graph is a graph on 2n vertices with "picture" as shown below:

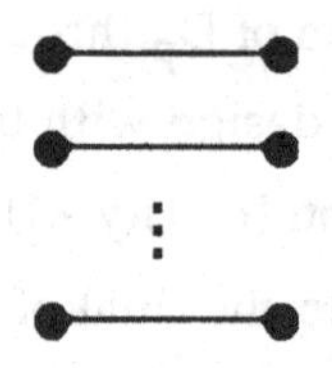

The complement of a ladder graph is called a cocktail party graph.

We mention (for later use) a result attributed to Wielandt (See Cameron and van Lint [49, p. 23]).

Theorem 3.19. Suppose Γ is a strongly regular graph on n = 2m vertices, whose eigenvalues have multiplicities 1, m - 1 and m. Then either

(i) Γ or its complement is a ladder graph, or

(ii) Γ or its complement has parameters

$n = 4s^2 + 4s + 2$, $a = s(2s + 1)$, $c = s^2 - 1$, $d = s^2$ for some positive integers.

We now give some early results of Goethals and Seidel [67]. The common theme in results of this type is to get structural or parametric information about a quasi-symmetric design D by assuming some extra structure on its block graph Γ.

Theorem 3.20. The only quasi-symmetric 2 - designs whose block graph Γ is a ladder graph are the designs consisting of two copies of a symmetric design.

Theorem 3.21. There is no quasi-symmetric block design whose block graph is $L_2(n)$ or its complement.

We remark that the original proofs of these above results use eigenvalue methods dealing with the $(0, \pm 1)$ adjacency matrix introduced by Seidel.

The following result is also due to Goethals and Seidel [67]. We give the proof from Cameron and van Lint [49, p. 30], since it also illustrates the use of some earlier mentioned results.

Theorem 3.22. A quasi-symmetric 2-design with $b = 2v - 2$ is either a Hadamard 3 - design or the unique 2-(6, 3, 2) design.

Proof. The eigenvalues of the block graph Γ are seen to have multiplicities $1, v - 1, b - v = v - 2$. So Theorem 3.19 applies. Also $v = 2k$, $b = 4k - 2$, $r = 2k - 1$. Take any block B and count flags (x, C), where $x \in B \cap C$, C a block $\neq$ B. We get $ay + (4k - 3 - a)x = 2k(k - 1)$. If Γ or its complement is a ladder graph, we may assume $a = n - 2$, and hence $4(k - 1)y + x = 2k(k - 1)$. Thus $2(k - 1)$ divides x; since $x \leq k$ we have $x = 0$, $y = k/2$. Now by Corollary 3.18, D is a 3-design and consequently a Hadamard 3-design. Otherwise, we have $v - 1 = 2s^2 + 2s + 1$, $k = s^2 + s + 1$, $a = s(2s +1)$ and so $s(2s + 1)y + (s + 1)(2s + 1)s = 2s(s + 1)(s^2 + s + 1)$. Thus $2s + 1$ divides $2s(s + 1)(s^2 + s + 1)$ which implies $s = 1$. It can be easily seen that in this case D is a unique design with parameters 2 - (6, 3, 2) which is the other conclusion in the statement of the theorem.

The following simple lemma is another useful tool in studying quasi-symmetric designs.

Lemma 3.23. In any quasi-symmetric 2-(v, k, λ) design D with intersection numbers x and y, the following relations hold:

(i) $k(r - 1)(x + y - 1) + xy(1 - b) = k(k - 1)(\lambda - 1)$.

(ii) $b = 2a - d + \bar{c} + 2$. Here, a and d are the usual parameters of the block graph Γ and $\bar{c}$ is the number of triangles on an edge of $\bar{\Gamma}$, the complement of Γ.

(iii) If $x = 0$ then $(r-1)(y-1) = (k-1)(\lambda-1)$.

Proof. Take any fixed block B of D. There are a blocks meeting B in y points and the remaining $b - 1 - a$ blocks meeting B in x points. Thus, counting flags (p, C), where C is a block $\neq$ B, and $p \in B \cap C$, we get

$$ay + (b - 1 - a)x = k(r - 1).$$

Next, count (p, q, C), where C is a block $\neq$ B and $p, q \in B \cap C$ to obtain $ay(y - 1) + (b - 1 - a)x(x - 1) = k(k - 1)(\lambda - 1)$.

Then (i) follows easily. Next, let B_1, B_2 be two non-adjacent vertices in the block graph Γ. Count the occurrences of the remaining blocks in two ways to obtain (ii). Finally, (iii) follows from (i) after the substitution $x = 0$.

Exercise 3.24. Show that the following relations hold in a quasi-symmetric design:

(i) $(r - \lambda)k(k - 1) = (y - x)(k - y - x)a + (b - 1)x(k - x)$.

(ii) $(b - 2r + \lambda)k(k - 1) = (k - x)(k - x - 1)(b - 1) - a(y - x)(2k - y - x - 1)$.

Also, use these equations to prove Lemma 3.23.

Hint: Use the matrix equation in the second proof of Theorem 3.11. Use the fact that the "all one" vector is a common eigenvector and substitute $(\in_0, \in_1) = (0, 1)$ and $(1, 0)$ respectively.

The next result of Tonchev [175, p. 153] is obtained by counting and carefully analyzing the possibilities. Our proof is somewhat different and the assertion is slightly stronger.

Theorem 3.25. In a 2-(v, k, λ) design D, $b \geq k(r - 1) - \binom{k}{2}(\lambda - 1) + 1$.

Equality holds if and only if any two blocks of D intersect in one or two points. In case of equality one of the following holds:

(i) D is a projective plane, i.e., $\lambda = 1$.

(ii) D is a biplane, i.e., D is symmetric and $\lambda = 2$.

(iii) D is residual of a biplane.

(iv) D is the complete 2–(5, 3, 3) design.

(v) $k = 2$ and D is the union of λ copies of K_3.

Proof. Let B be a fixed block and n_i = the number of blocks other than B. Then from the standard flag counting equations $\Sigma n_i = b - 1$, $\Sigma i n_i = k(r - 1)$ and $\Sigma \binom{i}{2} n_i = \binom{k}{2}(\lambda - 1)$ we get (by subtracting the second equation from the sum of the other two):

$$(b - 1) - k(r - 1) + \binom{k}{2}(\lambda - 1) = n_0 + \sum_{i=3}^{k} \binom{i-1}{2} n_i \geq 0$$

which proves the inequality and also shows that equality holds if and only if $n_i = 0$ except n_1 and n_2. Let equality hold. Then use of the standard parameter relations gives:

$$(*) \qquad r^2 - r[\lambda(k + 1)] + [\lambda k + \frac{k^2 \lambda(\lambda-1)}{2}] = 0$$

The discriminant of the above quadratic is $\Delta = \lambda[(k - 1)^2 - (\lambda - 1)(k^2 - 2k - 1)]$. If $k = 2$, then $\Delta = \lambda^2$ and $r = \lambda$ or 2λ, where the first possibility is ruled out since $v \neq k$ and the second gives $v = 3$ and yields case (v). If $k \geq 3$ and $\lambda \geq 3$, then $\lambda = k = 3$ (since $\Delta \geq 0$) and $r = 6$, $v = 5$ which yields case (iv). If $\lambda = 1$, then the quadratic gives the result that r divides k, which by Fisher's inequality yields $r = k$ and hence case (i). If $\lambda = 2$, then the two roots of (*) are $r = k$ and $r = k + 2$. The first gives case (ii) and the second gives (iii) after application of the Hall-Connor Theorem (Theorem 1.17).

Remark 3.26. Observe that while the inequality given in Theorem 3.25 is interesting, it should not be compared to Fisher's inequality since the former involves an expression in b, r, k, and λ (and b can be uniquely computed using r, k and λ) while the latter involves only b and v (which have no relation in general). For example, if $r - 1$ is less than $\frac{(k - 1)(\lambda - 1)}{2}$, then Theorem 3.25 gives no information.

Remark 3.27. Quasi-symmetric 4-designs have been completely determined by Ito et al. [98, 64] and Bremner [31]. In fact, the only non-trivial ones are the 4-(23, 7, 1) design or its complement. This deep number theoretic result, which in turn is based on a theorem of Ray-Chaudhuri and Wilson [129], will be discussed in later chapters after an in depth study of the 4-(23, 7, 1) design. We also note that any 5-design is a 4-design. Therefore, for the present it will suffice to conclude that in view of these observations, the existence problem of quasi-symmetric designs is open only in case of $t = 2$ and 3. This has also been mentioned as an open and difficult problem in the survey papers of Hedayat and Kageyama [78, 100] . This will be the subject of some later chapters in this monograph (see, e.g., Chapters VIII and IX).

For the sake of completeness, we conclude with the following result of Stanton and Kalbfleisch [166]. Their definition of quasi-symmetric designs was more restrictive than the general definition given at the beginning of this chapter.

Theorem 3.28. Let D be a quasi-symmetric 2-(v, k, λ) design with the property that there is precisely one block intersecting any given block in x (or y) points, i.e., in the block graph Γ or $\overline{\Gamma}$, $a = 1$ or $\bar{a} = 1$. Then either $x = 0$ and D is a Hadamard 3-design (with $y = \frac{k}{2}$) or $x = \lambda$, $y = k$ and D is obtained by taking two copies of a symmetric design.

Exercise 3.29. Use Theorem 3.19 to prove Theorem 3.28.

Let D be any quasi-symmetric 2-(v, k, λ) design with intersection numbers x and y with $x < y$. Let f and f' be the frequencies of y and x respectively. Put $s = \min\{f, f'\}$. Then Carmony and Tan [53] call D s-quasi-symmetric. Theorem 3.28 above deals with the case $s = 1$. Carmony and Tan consider designs which they denote by B(y, n, t). These are designs with $v = \frac{n^2y}{t^2}$, $b = \frac{n}{t^2}\frac{(n^2y-t^2)}{(n-1)}$, $r = \frac{(n^2y-t^2)}{(n-1)t}$, $k = \frac{ny}{t}$, $\lambda = \frac{(ny-t)}{(n-1)}$. Observe that the families B(y, n, t) have the parameters of an s-quasi-symmetric design with $s = n-1$, $x = \frac{ny(t-1)}{(n-1)}$. Carmony and Tan essentially give an algorithm which for any specific value of s determines the possible parameter sets of all potential s-quasi-symmetric designs. Moreover, a computer search in [53] for $s \le 15$ revealed certain exceptional parameters which do not belong to any B(y, n, t). The following result of M.S. Shrikhande and Singhi [156] shows that for any fixed s, there are no exceptional parameters for sufficiently large v.

Theorem 3.30. Let $s \geq 1$ be a fixed integer. Then any s-quasi-symmetric design 2-(v, k, λ) with intersection numbers x,y $(0 \leq x < y)$ satisfies one of the following conditions:

(i) $v < s^2 + s + 1$

or (ii) D is strongly resolvable with the parameters of $B(y, n, t)$, with $n = s + 1, 1 \leq t \leq s$.

IV. SOME CONFIGURATIONS RELATED TO STRONGLY REGULAR GRAPHS AND QUASI-SYMMETRIC DESIGNS

If D is a quasi-symmetric design, then we know from Chapter III that its block graph is strongly regular. We have also seen that by imposing some extra structure on the block graph we can pin down the design D. To a quasi-symmetric design, we can associate some other naturally associated strongly regular graphs. Conversely, starting from a graph we may sometimes produce a design. If the graph Γ has some further structure, then in some cases the associated designs may have additional properties. We also enlarge the class of incidence structures under consideration and study graphs associated with partially balanced incomplete block designs with a two class association scheme. Partially balanced incomplete block designs are simply 1-designs with occurrences of point-pairs in blocks determined by adjacencies in the superposed strongly regular graph.

A restricted class of the partially balanced incomplete block designs (PBIBDs) is the class of special partially balanced incomplete block designs (SPBIBDs) which, in addition, satisfy a point-block regularity condition. A generalization of the notion of an SPBIBD is the notion of a partial geometric design of Bose, S.S. Shrikhande, and Singhi [30]. The former structures were introduced by Bridges and M.S. Shrikhande [32] in order to obtain a unified approach to many graph embedding problems considered earlier. A nice outcome of the linear algebraic techniques is the spectral characterization of partial geometric designs by Bose, Bridges, and M.S. Shrikhande [23]. Some of these results (and their generalizations) form a few of the principal tools of Neumaier's work [117] on a possible approach to classification of quasi-symmetric designs discussed in Chapter VIII.

The proof techniques in this chapter use both the matrix and matrix-free approach. In fact, in all the situations where the former is only notationally used in the original literature, we have provided proofs using purely combinatorial arguments. An important example is the combinatorial characterization of the strongly regular graphs associated with the classical generalized quadrangle or a partial geometry $(q^2 + 1, q + 1, 1)$ due to Bose and S.S. Shrikhande [29]. On the other hand, our treatment includes proofs of Bose-Connor property for semiregular designs [24] using both the methods.

Convention 4.0 Before we begin, we caution the reader that the standard convention on the intersection number of a quasi-symmetric design will not (necessarily) hold in this chapter. In fact, the block graph (only for this chapter) is defined in such a way that two blocks are adjacent if and only if they intersect in x points (where $x \neq y$ but not necessarily $x < y$). This is convenient, since in most of the situations of interest in this chapter x will turn out to the the smaller intersection number. However, we make no such prior assumption.

Proposition 4.1. [49, p.38] Let Γ be a strongly regular graph with $c = 0$ and $1 < d < a$. Let p be any vertex of Γ and suppose $D(\Gamma, p)$ is the incidence structure whose points are the vertices adjacent to p and whose blocks are the vertices not adjacent to p. The incidence is defined by adjacency, i.e., a point-block pair is incident if the corresponding vertices are adjacent. Then $D(\Gamma, p)$ is a 2-design with $v = a$, $k = d$, $\lambda = k - 1$, $r = k - 1$ and $b = \frac{v(v - 1)}{k}$, with possibly repeated blocks.

Proof. Let α, β be any two points of $D = D(\Gamma, p)$. Since $c = 0$, the d vertices adjacent to both α and β are in the set consisting of p and the set of vertices not adjacent to p, i.e., $\lambda = d - 1$. For a vertex α not adjacent to p, α is certainly adjacent to $k = d$ vertices in the neighborhood of p. Clearly $v = a$ and hence the proof.

Theorem 4.2. (Biggs [16]) Let Γ be a strongly regular (SR) graph with degree of regularity $a = v$, $c = 0$ and $d = k$ where $d \neq a$ (as in Proposition 4.1) such that Γ has at least three vertices. Then the following assertions hold.

(i) $v = (s + 1)k + s^2$ where s is some positive integer.

(ii) k divides $s^2(s^2 - 1)$ and $k + 2s$ divides $s(s + 1)(s + 2)(s + 3)$.

(iii) For any $k \neq 2, 4, 6$ the number of SR graphs Γ with $c = 0$ is finite.

Proof. Let n denote the number of vertices of Γ. The case $k = d = 1$ is left to the reader. If $k = d \geq 2$, then Theorem 2.23 (ii) is applicable and

$$\frac{1}{2}\left[(n - 1) \pm \frac{A}{B}\right] \text{ are integers where}$$

$A = v(v + k - 3)$ (use $n - 1 = b + v$ and Proposition 4.1) and $B = \sqrt{k^2 - 4k + 4v}$. Hence, $k^2 + 4(v - k)$ is a perfect square and since $v \neq k$, $k^2 + 4(v - k) = (k + 2s)^2$ for some positive integer s. Therefore, $v = (s + 1)k + s^2$ which proves (i). Observe that Proposition 4.1 shows that the number of vertices not adjacent to p is $b = \frac{v(v-1)}{k}$ and hence k divides $v(v - 1) \equiv s^2(s^2 - 1) \pmod{k}$. Clearly, $B = k + 2s$ divides $A = v(v + k - 3)$, i.e., $k + 2s$ divides $s(s + 1)(s + 2)(s + 3)$ which implies (ii). Therefore, $k + 2s$ divides $k(k - 2)(k - 4)(k - 6)$. If k is fixed and $k \neq 2, 4,$ or 6, then s has only finitely many possibilities. Thus v and b have only finitely many possibilities and we are done because $v + b = n - 1$.

Remarks 4.3. The procedure outlined in Proposition 4.1 is sometimes reversible and [16] contains a discussion of the situation of Higman-Sims graph. We will also return to this point in Chapter VI. The theme of finding designs inside an SR graph is the main idea behind many theorems of the present chapter and will be elaborated later on. By using the well known upper bound (due to Majumdar, see [108]) on the number of blocks disjoint from a given block in a design, the following can be proved.

Theorem 4.4. Let Γ be an SR graph with $c = 0$ and $D = D(\Gamma, p)$ as in Proposition 4.1. Then the number of points of $D = v \geq \frac{1}{2}[3k + 1 + (k - 1)\sqrt{4k + 1}]$ with equality if and only if D is quasi-symmetric with $x = 0$.

Theorem 4.5. Let Γ be an SR graph with $c = 0$ and let $D = D(\Gamma, p)$ be as in Proposition 4.1. Then the following are equivalent:

(i) D is a 3-design.

(ii) D is quasi-symmetric with $x = 0$.

(iii) $\Gamma | \overline{\Gamma(p)}$ is strongly regular where $\overline{\Gamma(p)}$ is the set of vertices not adjacent to p.

(iv) $v = \frac{1}{2}[(3k + 1) + (k - 1)\sqrt{4k + 1}]$.

Further, any one of these conditions implies that $v = s(s^2 + 3s + 1)$ and $k = s^2 + s$.

Proof. Theorem 4.4 has already claimed equivalence of (ii) and (iv). By Proposition 4.1, $b = \frac{v(v-1)}{k}$. Hence by Corollary 3.18, (i) and (ii) are equivalent. By Cameron's Theorem 1.29, any one of (i) and (ii) gives four possibilities for D. Out of these, the rationality conditions and Theorem 4.4 leave us with $v = y(y^2 + 3y + 1)$, $k = y^2 + y$ where y is the positive block intersection number. Since $\Gamma \mid \overline{\Gamma}(p)$ is just the block graph of D, (iii) is easily verified. Let (iii) hold. Then the construction of D tells us that any block-pair has only two possible block intersection numbers. Since Γ has $c = 0$, one of these possibilities is $x = 0$ (justify this!). Hence (iii) implies (ii).

We now embark upon a related but somewhat different theme. Let D be a quasi-symmetric design with incidence matrix N and adjacency matrix A. When does the matrix

$$\begin{pmatrix} O & N \\ N^t & A \end{pmatrix}$$

represent an adjacency matrix of a strongly regular graph? Observe first that the question makes sense because the matrix is symmetric with zero diagonal. In other words, form a graph G whose vertices are the v points and b blocks of D. Two points are never adjacent, a point-block pair is adjacent if incident (compare Proposition 4.1), and a block-block pair is adjacent if it is adjacent in the block graph (i.e., intersects in x points by Convention 4.0). When does G become strongly regular? This question was considered by M.S. Shrikhande [145].

Theorem 4.6. Let D be a quasi-symmetric $(v, b, r, k, \lambda; x, y)$ design. Then the graph G defined above is strongly regular if and only if the parameters of D are:

$$v = \frac{k^2 + kg - g^2 - g}{k - g^2}, \lambda = \frac{k(k - g^2)}{g + 1}$$

$x = k - g^2 - g$, $y = k - g^2$, where $1 \le g \le \left(\frac{-1 + \sqrt{1 + 4k}}{2}\right)$, (i.e., D is a two parameter family). Moreover, in that case, $a^* = r = \frac{k(k - g)}{g + 1}$, $c^* = g + \lambda - k$, and $d^* = \lambda$, where *'s refer to the parameters of G.

Corollary 4.7. [161] Suppose D is a quasi-symmetric design with $x = 0$ and G is formed as above. Then G is strongly regular if and only if $v = (g + 1)(g^2 + 2g - 1)$, $k = g(g + 1)$, $\lambda = g^2$, and $y = g$.

We note that Corollary 4.7 is a special case of Theorem 4.6, a comparison with Theorem 1.29 shows that G has parameters of the block graph of a 3-design with $x = 0$ and D has parameters of a quasi-residual of such a 3-design. Whether this quasi-residual is actually a residual or not is a difficult question in general and will be considered in Chapter VII.

Proof of Theorem 4.6. Let H be the block graph of D (defined using intersection number x as per Convention 4.0). Denote the parameters of H by b, a, c, d and those of G by n^*, a^*, c^* and d^*. We then have $a^* = a + k$ and $a^* = r$ (counting for blocks and points respectively). Hence $a = r - k$. Since no two points are adjacent, $d^* = \lambda$ and for two non-adjacent blocks, the number of points (respectively blocks) adjacent to both is y (respectively d). So we have $d^* = \lambda = y + d = y + a + \theta_1$ and θ_2.

Using Exercise 3.11, $\theta_1\theta_2 = -\frac{(r - \lambda - k + y)(k - y)}{(y - x)^2}$ and $a = r - k$ was just obtained. Therefore, elimination of the graph parameters gives us $(r - \lambda - k + y)[(y - x)^2 - (k - y)] = 0$, where the first factor equals zero if and only if $\theta_1\theta_2 = 0$ which is forbidden by Exercise 3.12. Hence $k - y = (y - x)^2 = g^2$ where g is some positive integer.

Case 1 $y > x$. Then $k = y + g^2$ and $x = k - g^2 - g$. We now count the parameter of $d^* = \lambda$ in a different way. Fix a non-flag (p, B) and count the number of blocks on p meeting B in x points to get $d^* = \frac{yr - k\lambda}{y - x} = \frac{yr - k\lambda}{g}$. Hence $(k - g^2)r = (k + g)\lambda$. Finally, Exercise 3.11 tells us that $\frac{y(b - 1) - k(r - 1)}{g} = a = r - k$ which simplifies to $b(k - g^2) = (k + g)(r - g)$. Since $\frac{r}{\lambda} = \frac{v - 1}{k - 1}$, the first equation determines v uniquely in terms of k and g while the second equation divided by the first determines r in terms of k and g.

Clearly, then b is uniquely determined and hence the result is true.

Case 2 $x > y$. Then $y = k - g^2$ and $x = k - g^2 + g$. All the computations of Case 1 go through to yield $b(k - g^2) = (r + g)(k - g)$ and $r(k - g^2) = \lambda(k - g^2)$. Hence $\frac{v}{k} = \frac{b}{r} = \frac{r + g}{\lambda} \geq \frac{r}{\lambda} = \frac{v - 1}{k - 1}$ implies $v = k$, a contradiction.

Exercise 4.8. Let G be an SR graph containing an SR subgraph H with adjacency matrices P and A respectively. Suppose P can be written in the form

$$P = \begin{pmatrix} O & N \\ N^t & A \end{pmatrix}.$$

Show that in this case the parameters of G are as described in Theorem 4.6 and G arises from a quasi-symmetric D with incidence matrix N. (Hint: It suffices to show that N has two column inner products).

We briefly discuss the question of existence of designs in Theorem 4.6 and the corresponding SR graphs. The rationality conditions (Theorem 2.23) give no additional information. For $g = 1$, the designs D trivially exist and are complements of the designs on edges of a complete graph K_{k+2} (which corresponds to a triangular graph). Note also that with $k = 2$ and 3, the corresponding graph G is the Petersen graph and the (15, 6, 1, 3) graph respectively. Let $g = 2$. We get the (21, 6, 4) quasi-symmetric design (with $k = 6$) which has been introduced in Chapter III and will be discussed later. For $k = 7$, we get a design whose block graph is a (57, 14, 1, 4) SR graph and such a graph does not exist by a result of Wilbrink and Brouwer [179]. For $k = 13$, we obtain a (21, 105, 65, 13, 39) design with $x = 7$ and $y = 9$. This design can be ruled out using a result of Calderbank discussed in a later chapter. We leave it to the reader to prove that for a fixed $g \geq 2$, <u>there are finitely many designs satisfying the conditions of Theorem 4.6</u>.

We now turn our attention to an important class of strongly regular graphs studied by Mesner [112]. These graphs, the so called negative latin square graphs $NL_g(n)$, are a two-parameter family with $v = n^2$, $a = g(n + 1)$, $c = (g + 1)(g + 2) - (n + 2)$, and $d = g(g + 1)$. A construction of $NL_g(g^2 + 3g)$ was given by Mesner and we motivate the topic by asking the reader to work out the following:

Exercise 4.9. Let D be a quasi-symmetric 3-design with parameters $v = g(g^2 + 3g + 1)$, $k = g(g + 1)$, $\lambda_3 = g - 1$, $x = 0$, and $y = g$ (compare Theorem 1.29). Form a block graph H of D by making two blocks adjacent if they are disjoint. Compute the parameters of this SR graph H.

(Hint. Use Theorem 4.5 and/or Theorem 1.29).

Exercise 4.10. Let D be as in Exercise 4.9. Form a new graph G whose vertex set comprises the point-set and the block-set of D and a new symbol ∞ which is adjacent to all the points of D, the point-block and the block-block adjacency as defined exactly in Theorem 4.6, and two points as in that case are never adjacent. Show that G is an $NL_g(g^2 + 3g)$ containing H as a subgraph.

Observe that an $NL_g(g^2 + 3g)$ graph G has the "c" parameter equal to 0, i.e., G contains no triangles. This observation is crucial to the proof of the following result of Mesner [112].

Theorem 4.11. Let G be an $NL_g(g^2 + 3g)$. Suppose ∞ is any (distinguished) vertex of G. Define an incidence structure D whose points are vertices in the neighborhood of ∞ and where blocks are the other vertices (as in Proposition 4.1). Then D is a quasi-symmetric 3-design with $x = 0$, $v = g(g^2 + 3g + 1)$, $k = g(g + 1)$, $\lambda_3 = g - 1$, and $y = g$.

Proof. This is a direct consequence of Theorem 4.5. However, we give the following counting argument for the sake of variety.

Since G is triangle-free, Proposition 4.1 implies that D is a design with parameters $\lambda_2 = g^2 + g - 1$ and v, k as given in the statement of the Theorem. Let B be any block. The vertices of G adjacent to B correspond to the points incident with B and blocks disjoint from B. Since $a = g(g^2 + 3g + 1)$, the number of blocks disjoint from B is $g^3 + 2g^2$. But the total number of blocks of D is $b = (g^2 + 2g - 1)(g^2 + 3g + 1)$.

Hence there are $(g + 2)^2 g^2 - (g + 2) = (g^3 + 2g^2 - 1)(g + 2)$ blocks that have non-empty intersection with B. Let a_i be the number of those blocks that intersect B in i points, $i \geq 1$. Then $\Sigma a_i = (g^3 + 2g^2 - 1)(g + 2)$, $\Sigma i a_i = k(r - 1) = g(g + 1)(g^3 + 3g^2 + g - 2)$, and $\Sigma i(i - 1)a_i = k(k - 1)(\lambda_2 - 1) = g(g + 1)(g^2 + g - 1)(g + 2)(g - 1)$. From these three equations, $\Sigma(i - g)^2 a_i = 0$. Therefore every block intersects B in 0 or $y = g$ points, i.e., D is a quasi-symmetric design. Note that $b = \frac{v(v - 1)}{k}$ and invoke Corollary 3.18 to complete the proof.

Now consider a general situation. Let D be a $(v, b, r, k, \lambda ; x, y)$ quasi-symmetric design and let ∞ be a 'new' vertex. Make a graph G with vertex-set consisting of the point-set, the block-set, and ∞. Then G has $v + b + 1$ vertices. The following table represents the adjacencies in G:

∞ - point	adjacent
point - point	not adjacent
point - block	adjacency by incidence
block - block	adjacency in block graph

In other words, G is a graph with adjacency matrix P where

$$P = \begin{bmatrix} 0 & 1_v & 0_b \\ 1_v^t & O & N \\ 0_b^t & N^t & A \end{bmatrix}$$

and where $0 = 0_{v \times v}$, 1_v is the all 1 row vector of order v, 0_b is the all 0 row vector of order b, and A is the adjacency matrix of the block graph of D (this situation is converse of the one in Theorem 4.5 and also without an extra assumption on the "c" parameter of G). When is G strongly regular? The answer is given by

Theorem 4.12 [145] With everything as described above, G is a strongly regular graph if and only if D is a quasi-symmetric 3-design with $v = g(g^2 + 3g + 1)$, $k = g(g + 1)$, $\lambda_3 = g - 1$, $x = 0$, and $y = g$. In that case G is an $NL_g(g^2 + 3g)$.

Proof. The second assertion follows from Mesner's results. Let G be an SR graph. We denote the parameters of G by n^*, a^*, c^*, d^* and those of H, the block graph of D, by b, a, c, d as in the proof of Theorem 4.6.

No two points are adjacent in G and hence G has no triangle containing ∞. So $c^* = 0$. If two blocks are commonly incident with x points, then they are adjacent and $c^* = 0$ forces them not to have any common point. Hence $x = 0$. Our proof is based on three-way counting of the parameter d^*. For two points (which are not adjacent), the number of blocks commonly adjacent is λ and therefore $d^* = \lambda + 1$

(including ∞). Also, ∞ and a block are commonly adjacent to k points. So $d^* = k$. Finally, let (p, B) be a non-flag. How many blocks on p are disjoint from B? Since $x = 0$, these are easily seen to be $r - \frac{k\lambda}{y}$ in number. Therefore, $d^* = r - \frac{k\lambda}{y} = k$. But $\lambda = k - 1$ and from Lemma 3.23 (iii), $(r - 1)(y - 1) = (k - 1)(\lambda - 1) = (k - 1)(k - 2)$. Substitution of $r = k + \frac{k\lambda}{y} = k + \frac{k(k-1)}{y}$ then gives the quadratic $k^2 - (y^2 + y + 1)k + (y^2 + y) = 0$ whose roots are $k = 1$ or $k = y^2 + y$, where the first is absurd. So $k = y^2 + y$. Since $x = 0$, any derived design D_p of D is a 1-design with equally many points and blocks (find v and r using $\lambda = k - 1$, $k = y^2 + y$ and $(r - 1)(y - 1) = (k - 1)(\lambda - 1)$). But any two blocks of D_p intersect in $y - 1$ points, i.e., D_p is the dual of a design which is symmetric since it has equally many points and blocks (because $r = v - 1$). Thus D_p is a symmetric design, i.e., D is a 3- design with $x = 0$ (by Theorem 1.29) as asserted.

Essentially, Theorem 4.12 and Theorem 4.5 are equivalent. We leave the proof of this rather obvious assertion to the reader as an exercise. This makes the above proof unnecessary.

At this stage of development, we can no longer proceed without the introduction of partially balanced incomplete block designs. To that end we make the following definition.

Definition 4.13. An association scheme with t associate classes is a set S whose set of 2-subsets is partitioned into t subsets (association classes) and satisfies the following properties: For two elements p, q of S we say that p and q are i-associates of each other if the subset {p, q} is in the i-th association class. Then there exist constants n_i, P^i_{jk}, i, j, k = 1, ..., t such that the following conditions hold (for all i, j, k).

(i) For any element p of S , the number of elements q

which are i-associated with p is n_i.

(ii) For any two elements p and q which are i-associated, the number of elements which are j-associated to p and k-associated to q is P^i_{jk}.

In other words, if a complete graph with vertex set S is edge-colored in such a way that the i-th color is used for the i-th association class, then at each vertex the i-th color degree is n_i and for any edge with color i the number of triangles with other sides of colors j and k respectively (in that order) is P^i_{jk}. In fact, this interpretation proves the following:

Lemma 4.14. Let S be a 2-class association scheme on v elements. Then $n_1 + n_2 = v - 1$, $P^1_{11} + P^1_{12} = n_1 - 1$, $P^1_{12} = P^1_{21}$, $P^2_{11} + P^2_{12} = n_1$. Hence the parameters v, n_1, P^1_{11} and P^2_{11} determine all the other parameters uniquely.

Exercise 4.15. Let G be (v, a, c, d) strongly regular graph. Define two vertices to be first (respectively second) associates if they are adjacent (respectively not adjacent) in G. Show that this defines a 2-class association scheme with $n_1 = a$, $P^1_{11} = c$, and $P^2_{11} = d$. Show that this procedure can also be reversed to obtain an SR graph from a 2-class association scheme.

Association schemes with two associate classes are sufficient for the discussion to follow. However, we note that the notion of strongly regular graphs was implicit in the work of Bose and Nair [26] and Bose-Shimamoto [27], and historically association schemes were formally defined much before strongly regular graphs were defined by Bose [21] in 1963. The book by Raghavarao [127] describes 2-class schemes in detail and the book by Bannai and Ito [11] is devoted to the algebraic study of association schemes. The significance of association schemes in algebraic combinatorics and coding theory was perhaps first recognized by Delsarte [60]. As already asserted strongly regular graphs are the same as 2-class association schemes and in our treatment we prefer to put everything in the language of the former.

Definition 4.16. A partially balanced incomplete block design (PBIBD) (with 2-class scheme) is a (v, b, r, k) 1-design D with the following additional property: There is a strongly regular graph on the points of D (to be called the point-graph of D), say G = G (D), and constants λ_1 and λ_2 such that two points are adjacent (respectively not adjacent) in G if and only if they are contained in λ_1 (respectively in λ_2) blocks.

Essentially, a PBIBD is a 1-design with a superposed point-graph which determines the occurrence of a point-pair in the blocks. We note that a design can be considered as a PBIBD with G either a complete graph or its complement. A detailed discussion of PBIBDs is beyond the scope of this book; we merely refer to Raghavarao [127] or Beth, Jungnickel, and Lenz [15]. The following is an important class of PBIBDs.

Definition 4.17. A group divisible design (GDD) is a PBIBD with v = mn points satisfying the property that the point-graph G = G(D) is a vertex disjoint union of m complete subgraphs each of size n. These subgraphs are called groups (without confusing with the same term used in algebra!).

Clearly, the point-graph G of a group divisible design has parameters v = mn, a = n - 1, c = n - 2 and d = 0. If N denotes the incidence matrix and A the adjacency matrix of the point-graph of a

GDD, then a simple counting argument shows that $NN^t = rI + \lambda_1 A + \lambda_2 (J - I - A)$. But the point-graph G is a union of complete graphs and hence $A = \mathrm{diag}(J - I, \ldots, J - I)$ where each constituent has order n. From this, it is easily seen that the eigenvalues of NN^t other than the row sum are $r-\lambda_1$ and $rk-\lambda_2 v$. Though there are other types of GDDs the one we need here is

Definition 4.18. A GDD is called semiregular (SRGDD) if $rk = \lambda_2 v$ and $r > \lambda_1$.

The following property of SRGDDs is basic to all the discussion to follow.

Lemma 4.19. (Bose-Connor property [24]) If D is an SRGDD $(v, b, r, k, \lambda_1, \lambda_2, m, n)$, then every block contains exactly k/m points from each group. In other words, the incidence structure induced on every group is an $(n, k/m, \lambda_1)$ - design (which is possibly trivial).

Proof. Fix a group and let the i-th block contain a_i points of that group. Then $\sum a_i = nr$ and $\sum a_i(k-a_i) = n^2(m-1)\lambda_2$, hence $\sum a_i^2 = \lambda_2 n^2$. Therefore, $\sum (a_i - \frac{k}{m})^2 = 0$ follows easily after using $bk = vr$, $v = mn$, and $rk = \lambda_2 v$, since D is semiregular. Hence $a_i = k/m$ for all i.

Exercise 4.20. Show that if D is an SRGDD with $\lambda_1 = 0$, then every block contains one point from every group and $k = m$.

The family of SRGDDs is large and we have to refer the reader to the literature for details. We restrict ourselves to an important construction, basing ourselves on the following classical result.

Theorem 4.21. Let Q be a quadric in PG(3, q^2). The incidence structure whose points are the points of Q and whose blocks are those lines contained (fully) in Q forms a generalized quadrangle with $q^2 + 1$ points on every block and $q + 1$ blocks on each point, i.e., a partial geometry $(q + 1, q^2 + 1, 1)$.

For a proof, refer to [61]. We also record that this incidence structure has the property that if three points (lines) are pairwise collinear (concurrent), then they lie on the same line (are concurrent at a single point), i.e., we have no triangles. Since the dual of a partial geometry is also a partial geometry, we obtain two (geometric) graphs from the above generalized quadrangle. For example, with $q = 2$ the graph on lines is the unique (27, 10, 1, 5) graph called the <u>Schläfli graph</u> (see [49]) and the graph on points is the (45, 12, 3, 3) graph. We concentrate on the graph on the lines of this generalized quadrangle.

Exercise 4.22. Show that the graph G on the lines of the above generalized quadrangle has parameters $v = (q + 1)(q^3 + 1)$, $a = q(q^2 + 1)$, $c = q - 1$, and $d = q^2 + 1$ (as a strongly regular graph).

Bose and S.S. Shrikhande [29] call an SR graph G with the above parameters a pseudo-geometric $(q^2 + 1, q + 1, 1)$ - graph (the interchange of the first two parameters is due to the fact that graph of a partial geometry is its point graph by definition). For such a graph G they introduced two additional properties locally at a vertex ∞.

Property P G is said to have property P at a vertex ∞ if the set of vertices adjacent to ∞ can be partitioned into sets S_i, $i = 1, \ldots, q^2 + 1$, each S_i containing q vertices such that for every i, $T_i = \{\infty\} \cup S_i$ is a clique (of order $q + 1$).

Property P* G is said to have property P* at a vertex ∞ if G does not have the following as an induced subgraph.

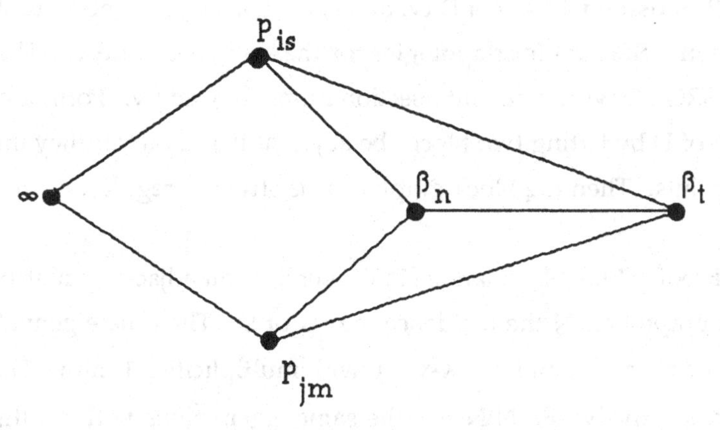

where $i \neq j$, $p_{is} \in S_i$, $p_{jm} \in S_j$.

In plain terms, if we begin with G coming from a generalized quadrangle then the lines p_{is} and p_{jm} intersect the line ∞ at distinct points and hence are not concurrent (else we get a triangle). Also the lines β_n, β_t, and p_{is} are pairwise intersecting and therefore must be concurrent at a point. The same argument applied to β_n, β_t, and p_{jm} shows that β_n, β_t, and p_{jm} are also concurrent, i.e., p_{is} and p_{jm} are concurrent, a contradiction.

Exercise 4.23. Show that a geometric graph $(q^2 + 1, q + 1, 1)$ has both the properties P and P* with respect to every vertex ∞.

Bose and S.S. Shrikhande [29] show that a pseudo-geometric graph G satisfying P and P* with respect to some vertex ∞ gives rise to an incidence structure which is an SRGDD which is also quasi-symmetric (i.e., has two block intersection numbers). They also prove that the block graph (in this case) is strongly regular. We begin by proving the following stronger version of the latter assertion.

Proposition 4.24. Let D $(v, b, r, k, m, n, \lambda_1, \lambda_2; x, y)$ be a quasi-symmetric SRGDD (our apologies for the symbol overflow). That is, D is an SRGDD with block intersection numbers x and y. Form a block graph of D by letting two blocks be adjacent if and only if they intersect in x points. Then the block graph of D is strongly regular.

Proof. (M.S. Shrikhande [148]) Let B be an adjacency matrix of the block graph and N the incidence matrix of D. Then the eigenvalues of NN^t are rk, $r - \lambda_1$, and $rk - \lambda_2 v = 0$ with multiplicities 1, m(n - 1) and m-1 respectively. So N^tN has the same eigenvalues with multiplicities 1, m(n - 1), and b - m(n - 1) -1 respectively. But $N^tN = kI + xB + y(J - I - B)$. Therefore, the eigenvalues of B are

$$\theta_0 = \frac{y(b-1) - k(r-1)}{y-x} \text{ with multiplicity } 1,$$

$$\theta_1 = \frac{(r-\lambda_1) - (k-y)}{y-x} \text{ with multiplicity } m(n-1),$$

and $$\theta_2 = \frac{k-y}{y-x} \text{ with multiplicity } b - m(n-1) - 1.$$

Hence B represents an SR graph by Theorem 2.22.

Theorem 4.25. (Bose and S.S. Shrikhande [29]) Let G be a pseudo-geometric graph - $(q^2 + 1, q + 1, 1)$ satisfying properties P and P* w.r.t. some vertex ∞. Define an incidence structure $D(\infty)$ whose blocks are the vertices not adjacent to ∞. The incidence in $D(\infty)$ is defined by adjacency in G. Then D (∞) is a quasi-symmetric SRGDD with parameters

$$v' = q(q^2 + 1),\ b' = q^4,\ r' = q^3,\ k' = q^2 + 1,$$

$$m' = q^2 + 1,\ n' = q,\ \lambda_1' = 0,\ \lambda_2' = q^2,\ x' = 1, \text{ and } y' = q + 1.$$

Proof. Using the parameters of G, it is easily seen that D = D(∞) is a 1-design with the given parameters v', b', r', k'. Let p_{is} and p_{im} be in S_i. Then property P and the fact that c = q - 1 implies that these two vertices are not commonly adjacent to any vertex not in the neighborhood of ∞. Essentially the structure of G is locally determined at ∞: $T_1, ..., T_{q2+1}$ are cliques of size q + 1 mutually intersecting in ∞ alone. It also follows that there are no edges between T_i and T_j for $i \neq j$. Therefore, if two points are in different sets S_i and S_j, then they commonly occur in d = $q^2 + 1$ blocks. Hence D is a GDD with m' = $q^2 + 1$, n' = q (and with $S_1, ..., S_{q2+1}$ as groups) such that $\lambda_1' = 0$ and λ_2' = q + 1. Since $r'k' = \lambda_2' v'$, D is an SRGDD. It remains to show that D is quasi-symmetric. This is done by proving that two blocks intersect in a single point if they are adjacent (in G) and intersect in q + 1 points otherwise, i.e., essentially the induced graph on the set of vertices not adjacent to ∞ is the block graph of D. Let B_n and B_t be two blocks such that the corresponding vertices are adjacent. Since S_i's are groups and $\lambda_1 = 0$, these blocks contain at most one point of a given S_i. But property P* also tells us that these two blocks cannot have two common points from different S_i's. So B_n and B_t have at most one point in common. Equivalently, there is at most one triangle B_nB_t p containing the edge B_nB_t such that p is adjacent to ∞. But the number of blocks B_t adjacent to B_n is a - d = $q(q^2 + 1) - (q^2 + 1) = (q^2 + 1)(q - 1)$, while for every p, the number of such B_t's is c (this follows since vertices of different S_i's are not adjacent) and the number of such p's is clearly d. Since $cd = (q^2 + 1)(q - 1)$, the equality of these numbers shows that B_n, B_t are on exactly one triangle $B_nB_t\,p$ where p is adjacent to ∞. Hence $x' = 1$.

We still have to show that $y' = q + 1$. For a given block B_n, the number of blocks not adjacent to B_n is exactly $(q^4 - 1) - (q - 1)(q^2 + 1) = q\,(q - 1)(q + 1)$. Suppose the i-th block (of this collection) contains a_i points of B_n. Then $\sum a_i = (q^2 + 1)(r - 1 - c)$ (justify this using property P) and $\sum a_i(a_i - 1) = (\lambda_2 - 1)m'(m' - 1)n'^2$, i.e., $\sum a_i = (q^2 + 1)(q^3 - q)$ and $\sum a_i\,(a_i - 1) = (q^2 + 1)q^2(q^2 - 1)$. Hence $\Sigma(a_i - (q + 1))^2 = 0$. So every block not adjacent to B_n contains $q + 1$ point of B_n, i.e., $y' = q + 1$. This completes our proof.

The procedure in Theorem 4.25 is reversible as shown in the following [29] and some parts of the proof are left to the reader as an exercise.

Theorem 4.26. Let D be a quasi-symmetric SRGDD with parameters as given in the statement of Theorem 4.25. Let ∞ be a new vertex and define a graph G with adjacencies defined as follows:

∞ - point	adjacent
point - point	adjacent if in the same group
point - block	adjacent if incident
block - block	adjacency in block graph

Then G is a pseudo-geometric - $(q^2 + 1, q + 1, 1)$ graph.

Proof. It is easily seen that the number of blocks adjacent to a given block (intersection one) is $(q^2 + 1)(q - 1)$. Hence a routine computation shows that G is regular of degree $a = v' = q(q^2 + 1) = r' + 1 + (n' - 1) = k' + (q^2 + 1)(q - 1)$. For a point p the number of vertices adjacent to p and ∞ is $n' - 1$ (vertices in the group of p) while for vertices p and s in the same group the number of vertices adjacent to both is $n' - 2 + 1 = n' - 1$. For a flag (p, B) the number of blocks meeting B in p alone is $q - 1$ (and B contains no other point in the group of p). Finally, for a block-pair meeting at a single point p, there are $q - 2$ blocks meeting both in a single point and p is certainly adjacent to both the blocks. We have thus proved that $c = q - 1$.

Exercise 4.27. Complete the proof of Theorem 4.26 by showing that $d = q^2 + 1$. Also, show that the block graph of D is a negative latin square graph $NL_{q-1}(q^2)$.

When is the graph G obtained in Theorem 4.26 a geometric graph? That is, what are the necessary and sufficient conditions for D, a quasi-symmetric SRGDD with the right parameters, to arise from a geometric graph $(q^2 + 1, q + 1, 1)$ in such a way that $D = D(\infty)$ for some vertex? In other words, when is it possible to define an incidence structure in a generalized quadrangle? The precise answer to this question was given in the following theorem of Cameron, Goethals, and Seidel [50]; the proof uses more sophisticated combinatorial techniques.

Theorem 4.28. Every pseudo-geometric $(q^2+1, q+1, 1)$ - graph is a geometric graph.

In [29], it was observed that the condition of Theorem 4.28 holds if for every point p of the quasi-symmetric SRGDD the following holds: For blocks B, C containing p define $B \sim C$ if $B \cap C = \{p\}$. Then $\sim$ is an equivalence relation.

Observe that the general outline of the procedure in Theorem 4.26 (given in the table) is similar to the one in Theorem 4.5, Theorem 4.6, and Theorem 4.12. Is the situation in Theorem 4.26 unique? Barring trivialities, the answer to that question is in the affirmative as shown in Theorem 4.30 of M.S. Shrikhande [148]. Consider also the following equivalent formulation. Let D be a quasi-symmetric SRGDD with N, C, and B representing the incidence matrix, adjacency matrix of the point graph, and adjacency matrix of the block graph respectively. Let

$$A = \begin{bmatrix} 0 & j_v & 0_b \\ j_v^t & C & N \\ 0_b^t & N^t & B \end{bmatrix}$$

be a square matrix of order $v + b + 1$ where j_v is a row vector of size v of all ones and 0_b is a row vector of size b all zeros. Note that $C = \text{diag}(J - I, \ldots, J - I)$. When does A represent an adjacency matrix of a strongly regular graph?

Exercise 4.29 Show that the above matrix set-up is equivalent to the 'incidence' set-up in the statement of Theorem 4.26.

Theorem 4.30. Let D be a $(v, b, r, k, m, n, \lambda_1, \lambda_2; x, y)$ quasi-symmetric SRGDD. Let the graph G be obtained as given in the table in the statement of Theorem 4.26. Then one of the following holds.

(i) $v = q(q^2 + 1)$, $b = q^4$, $r = q^3$, $k = q^2 + 1 = m$, $n = q$, $\lambda_1 = 0$, $\lambda_2 = q^2$, $x = 1$, $y = q + 1$, and in that case G is a pseudo-geometric $(q^2 + 1, q + 1, 1)$ - graph.

(ii) $v = 2n$, $b = n^2$, $k = 2$, D is obtained as a complete bipartite graph on 2n points (with $m = 2$) and in that case G is an $((n + 1)^2, n^2, (n - 1)^2, n(n - 1))$ - graph which is obtained as a complement of the lattice graph $L_2(n + 1)$ and is unique except when $n = 3$.

Proof. The last statement in (ii) is obvious, using Theorem 2.28. Let v^*, a^*, c^*, d^* denote the parameters of the graph G, b, a, c, d those of the block graph H of D, also clearly $v = mn$, $a' = n - 1$, $c' = n - 2$ and $d = 0$ those of the point-graph of D (since D is group divisible). Then a straightforward three-way counting of a^*, c^* and d^* produces

$$a^* = v \quad = n + r \quad = k + a \tag{1}$$

$$c^* = n - 1 \quad = (n - 1) + \lambda_1 \quad = x + c \tag{2}$$

$$d^* = k \quad = 1 + \lambda_2 \quad = y + d \tag{3}$$

where (2) implies $\lambda_1 = 0$. Since D is semiregular and $\lambda_1 = 0$, the Bose-Connor property implies that $k = m$ and every block contains one point from each group. Since $rk = \lambda_2 v$ we can write the parameters of D: $v = mn$, $b = n^2\lambda_2$, $r = n\lambda_2$, $k = m$, $\lambda_1 = 0$, and $\lambda_2 = k - 1 = m - 1$ by (3).

Let (p, B) be a flag and count c^*, the number of blocks containing p and intersecting B in x points, to get

$$c^* = \frac{(r - 1)(y - 1) - (k - 1)(\lambda_2 - 1)}{y - x} \tag{4}$$

The same procedure for a non-flag (p, B) produces

$$d^* = \frac{ry + (y - x) - (k - 1)\lambda_2}{y - x}. \qquad (5)$$

The standard eigenvalue computation of the block graph H (i.e., adjacency matrix B) shows that they are $\theta_1 = -\frac{r - \lambda_1 - (k - y)}{y - x} = -\frac{r - k + y}{y - x}$ and $\theta_2 = \frac{k - y}{y - x}$ (also see the proof of Proposition 4.24). But $c = a + \theta_1 + \theta_2 + \theta_1\theta_2$ by Theorem 2.22. Hence using (1), $r - k = a - n$ and by (2), $x = n - 1 - c$. So $r - k + x = a - 1 - c = -(\theta_1 + 1)(\theta_2 + 1) = \frac{(r - k + x)(k - x)}{(y - x)^2}$. It follows that either $r - k + x = 0$ or $k - x = (y - x)^2$. If $r - k + x = 0$, then $r = k - x$, i.e., $n\lambda_2 = \lambda_2 + 1 - x \leq \lambda_2 + 1$ and hence $n \leq 2$. Since D is a PBIBD, $n = 2$ and $\lambda_2 = 1$. But $k = \lambda_2 + 1$ implies $k = 2$ and then $r = n\lambda_2 = 2 = k - 1 = 2 - x$ implies $x = 0$. We leave it to the reader to show that this leads to the second possibility.
Hence $k - x = (y - x)^2 = g^2$ for some positive integer g.

Case 1: $y > x$. We can then assume that $y - x = g$, $k = x + g^2$, $y = x + g$. Then (3) and (5) give (simplify using $\lambda_2 = k - 1$)

$$g + \lambda_2 = ny \qquad (6)$$

while (2) and (4) give

$$g(n - 1) = (k - 1)[n(y - 1) - (\lambda_2 - 1)] - (y - 1) \qquad (7)$$

which using (6) finally gives

$$g(n - g) = (k - 1)(g - n).$$

Since g and $k - 1$ are both positive, $g = n = q$ (say). Then (6) gives $\lambda_2 = q(y - 1)$. But $\lambda_2 = k - 1$ and hence (7) implies $q(n - 1) = (q - 1)(y - 1)$, i.e., $q = y - 1$. So $y - 1 = q = g = y - x$, i. e., $x = 1$ and $y = q + 1$ which gives (i).

Case 2: $y < x$. We can then assume that $x - y = g$, i.e., $k = x + g^2$ and $y = x - g = k + g^2 - g$. Then (3) and (5) give

$$g + \lambda_2 = ny \qquad (6')$$

while (2) and (4) give

$$g(n - 1) = (k - 1)[(\lambda_2 - 1) - n(y - 1)] + (y - 1) \qquad (7')$$

which simplifies to (using (6')) $g(n + g) = (k - 1)(n + g)$.
Since $n + g$ is positive, $g = k - 1$. But then $k = g + 1 = x + g^2$. So $x = g + 1 - g^2 \leq 1$. Since $0 \leq y < x$, we have $x = 1$ and $y = 0$. Then $g + 1 = x + g^2 = 1 + g^2$ implies $g = 0$ or 1, $g = 0$ implies $k = 1$ which is absurd, and hence $g = 1$, i.e., $k = 2$. This gives rise to (ii).

Observe that the Bose-Connor property tells us that the incidence structure induced by an SRGDD in any of its groups is a design. In that sense SRGDDs are similar to designs. Also the same property tells us that SRGDDs satisfy the following point-block regularity condition: Given a point-block pair (p, B), the number of points adjacent to p (in the point-graph), i.e., those in the group of p and which are also contained in p, depends only on whether (p, B) is a flag or non-flag (in fact, these numbers are k/m and $(k/m) -1$ respectively). A design trivially satisfies this property (by treating a design as a GDD with singleton groups). This property is the basis of the following definition introduced by Bridges and M.S. Shrikhande [32].

Definition 4.31. Let D be a (v, b, r, k) 1-design and let (s, t) be a pair of non-negative integers. D is called a special partially balanced incomplete block design (SPBIBD) of type (s, t) if there are constants λ_1 and λ_2 with the following property: Any two points are contained in λ_1 or λ_2 blocks. If (p, B) is a point-block pair, then the number of points on B which occur with p in λ_1 blocks is s (respectively t) if (p, B) is a flag (respectively a non-flag).

Declare two points of an SPBIBD adjacent in a graph G = G(D) if they are contained in λ_1 blocks and not adjacent otherwise. Then G is a (point) graph of D. Whether or not D is a PBIBD (and hence the name) depends on whether or not G is strongly regular. In Theorem 4.32, we will show that D is actually a PBIBD. To that end define P to be the adjacency matrix of G. Observe also that a design is an SPBIBD where λ_2 is vacuous, $s = k - 1$, and $t = k$.

Write N to denote the incidence matrix of an SPBIBD and call $\alpha = s - t$ the index of D.

Finally, let $\theta = \frac{r - \lambda_2}{\lambda_2 - \lambda_1}$ and $g = \lambda_1 - \lambda_2$. Then a standard counting argument produces

$$NN^t = (r - \lambda_2)I + (\lambda_1 - \lambda_2)P + \lambda_2 J, JN = KJ \quad (1)$$

$$PN = (s - t)N + tJ \quad (2)$$

and (1) and (2) can be written in the concise forms:

$$NN^t = g(P - \theta I) + \lambda_2 J \quad (3)$$

$$PN = \alpha N + tJ. \quad (4)$$

Theorem 4.32. An SPBIBD is a PBIBD.

Proof. It suffices to prove that P carries a strongly regular graph. By Theorem 2.22, it is enough to show that P has at most two eigenvalues other than the row sum.

Write

$$n_1 = \frac{rk - \lambda_2 v}{g} + \theta. \quad (5)$$

Then $PJ = n_1 J$. Hence from (3) (and using (4)) we obtain

$$g(P - \theta I)(P - \alpha I) = (tr - \lambda_2 (n_1 - \alpha))J. \quad (6)$$

So P has at most two eigenvalues θ and α other than n_1, which completes our proof.

Writing Λ to denote the set of (distinct) eigenvalues of P, we see that $\Lambda(P)$ is contained in $\{n_1, \alpha, \theta\}$. The proof of Theorem 4.32

implicitly yields several equations involving (the multitudes of) parameters. Equating traces in (6) and line sums in (4) gives

$$\lambda_1 n_1 = sr, \qquad (n_1 - \alpha)k = tv. \qquad (7)$$

We can, of course, use the standard parameter relations $bk = vr$ and $n_1\lambda_1 + (v - 1 - n_1)\lambda_2 = r(k - 1)$ (counting flags (q, B) in two ways where p is a fixed point and (p,B) is a flag). These, in conjunction with (5) and (7) give

$$s = \frac{\lambda_1}{r}\left(\frac{rk - \lambda_2 v}{g} + \theta\right), t = \frac{k}{r}\,\frac{(r - \lambda_1)}{v - k}\left(\frac{rk - \lambda_2 v}{g} + \theta\right). \qquad (8)$$

Hence, the expressions on the R.H.S. of (8) must be integers. Also, (7) and (8) give

$$(n_1 + \alpha\theta)k = t(n_1 - \theta). \qquad (9)$$

Since $\Lambda(P)$ is contained in $\{n_1, \alpha, \theta\}$ and n_1 is simple, the multiplicity ρ of a can be found to be

$$\rho = \frac{rk(rk + \lambda_2 - r - sg - \lambda_2 k)}{(g\alpha + r - \lambda_2)(gt - \lambda_2 k)}. \qquad (10)$$

Hence the expression on the R.H.S. of (10) must be an integer.

Theorem 4.33. [32] Let D be a PBIBD with parameters v, b, r, k, λ_1, and λ_2 and with P as an adjacency matrix of the point graph. Let $PJ = n_1 J$ and suppose θ is defined as before. Then D is an SPBIBD if and only if one of the following holds:

(i) $\theta \in \Lambda(P) - \{n_1\}$ and P is connected.

(ii) $\theta = n_1$ and $|\Lambda(P)| = 2$.

Proof. Notice that P has only two eigenvalues if and only if $P = \text{diag}(J - I, ..., J - I)$. In that case, D is a GDD. The proof is based on showing that an equation of the type (4) holds if and only if either of (i) and (ii) is true. We illustrate this in case (i). Since P is connected and strongly regular, $\Lambda(P) = \{n_1, \theta, \alpha\}$, where $\theta \neq \alpha$ for some α and n_1 is the degree of regularity. Let $X = PN - \alpha N = (P - \alpha I)N$. Then $NN^t = g(P - \theta I) + \lambda_2 J$ by (3) and therefore $XX^t = (P - \alpha I)^2 [g(P - \theta I) + \lambda_2 J]$. From the polynomial of P, we obtain $(P - \alpha I)(P - \theta I) = (n_1 + \alpha\theta)J$ (the reader is asked to prove this in Exercise 4.34). Hence XX^t is a multiple of J. Since rank (X) = rank (XX^t), we obtain rank (X) = 1. Therefore all the other rows (columns) are multiples of the first row (the first column). This forces $X = tJ$, i.e., $PN = \alpha N + tJ$, and clearly α is an integer.

Exercise 4.34. Let A be an adjacency matrix of an SR graph G with a, θ_1, θ_2 as distinct eigenvalues and a as the degree of regularity. Show that $(A - \theta_1 I)(A - \theta_2 I) = (a + \theta_1\theta_2)J$.

Hint: Use Remark 2.6 and Theorem 2.22.

Exercise 4.35. Let D be a $(v, b, r, k, \lambda_1, \lambda_2)$ SPBIBD of index $\alpha = s - t$ and let P be the adjacency matrix. Define $Q = J - I - A$. Show that defining two points to be adjacent if they occur in λ_2 blocks makes D an SPIBD of index $-\alpha-1$ with adjacency matrix (of the point graph) equal to Q.

Corollary 4.36 Let D be a PBIBD which is not a design. Then D is an SPBIBD if and only if rank(N) < v (where N is the usual vxb incidence matrix).

Proof. Let D be special. Then it is enough to show that N has a zero eigenvalue. We have $NN^t = g(P - \theta I) + \lambda_2 J$ by (3) where θ is an eigenvalue of P by (6). If $\theta = n_1$, then the only other eigenvalue of P must be -1 (of multiplicity v - 1). In that case, $P = J - I$, a contradiction since D is not a design. So $\theta \neq n_1$ and the eigenvalue of J corresponding to θ is clearly 0. So NN^t has a zero eigenvalue. For the converse, complementing P if necessary (see Exercise 4.35), we may assume that P represents a regular and connected graph. Hence the zero eigenvalue of NN^t must arise from an eigenvalue θ of P where θ is given by (3). If $\theta = n_1$, then (as before) P represents a disconnected graph, a contradiction. Hence $\theta \neq n_1$ and we are done by Theorem 4.33.

Exercise 4.37 Prove the Bose-Connor property for an SRGDD. Hint: Show that D is special with $\alpha = -1$ and $t = k/m$.

We recall that a PBIBD with triangular (respectively L_2) association scheme is one whose point graph is a triangular (respectively lattice) graph. Using Theorem 4.33, Bridges and M.S. Shrikhande [32] gave necessary and sufficient conditions for a PBIBD with any one of these schemes to be an SPBIBD. We refer the reader to [32] for details. The same paper also gives the following characterization of an SRGDD.

Theorem 4.38. Let D be a PBIBD with P as the adjacency matrix of its point-graph. Assume that D is not a design. Then the following are equivalent.

(i) D is an SPBIDB of index -1.

(ii) D is an SPBIBD of index 0 w.r.t. J - I - P.

(iii) D is an SRGDD.

(iv) D is a GDD and is special with $r > \lambda_1$.

Proof. (i) and (ii) are clearly equivalent by Exercise 4.35, and (iii) implies (i) by the Bose-Connor property. We now show that (i) implies (iii). Since $\alpha = -1$, (9) implies that $t = k$ or $n_1 = \theta$. If $t = k$, then (7) implies $v = n_1 + 1$, i.e., $P = J - I$ and D is a design, a contradiction. So $n_1 = \theta$. Hence P has two eigenvalues, i.e., $P = \text{diag}(J - I, \ldots, J - I)$. Therefore D is a GDD. If $r = \lambda_1$, then (8) gives $t = 0$ and (7) implies $(n_1 + 1)k = 0$, a contradiction, so $r > \lambda_1$. Since NN^t has two non-zero eigenvalues (and D is group divisible), the zero eigenvalue must be $rk - \lambda_2 v$. So D is semiregular. Hence (iii) follows.

It is now enough to prove that (iv) implies (iii). Since D is group divisible, $P = \text{diag}(J - I, \ldots, J - I)$. So $\Lambda(P) = \{n_1, -1\}$. But D is special and so $\theta \in \Lambda(P)$. If $\theta = -1$ then $r = \lambda_1$, a contradiction. Hence $\theta = n_1$ and then (5) gives $rk = \lambda_2 v$, proving (iii).

Observe at this stage that implicit in the proof of Theorem 3.8 is an assertion, that the dual of a quasi-symmetric design D is a PBIBD whose point-graph is, in fact, the block graph of D (and for which λ_1, λ_2 equal x, y of D). A more general result is the following result in [32].

Theorem 4.39. The dual of a quasi-symmetric SPBIBD is an SPBIBD.

Proof. With the standard notations

$$N^tN = (k - y)I + (x - y)A + yJ,$$

where A is the adjacency matrix of the block graph.

Write $g' = x - y$, $\theta' = \frac{k - x}{y - x}$ to get

$$N^tN = g'(A - \theta I) + yJ$$

Multiply this equation on the left by N to get

$$NN^tN = g'(NA - \theta' N) + yrJ \qquad (11)$$

and using (3) we get

$$NN^tN = g(PN - \theta N) + \lambda_2 kJ$$

which using (4) gives

$$NN^tN = g(\alpha - \theta)N + (gt + \lambda_2 k)J. \qquad (12)$$

Equating the R.H.S. of (11) and (12) we get

$$NA = \alpha' N + t'J$$

where $\alpha' = \frac{g(\alpha - \theta)}{g'} + \theta'$ and $t' = \frac{\lambda_2 k - yr + gt}{g'}$.

Thus $AN^t = \alpha' N^t + t' J$ (of appropriate orders). Comparison of the last equation with (4) and the basic definition of an SPBIBD show that the dual is an SPBIBD of index α'. This completes the proof.

In the general setting of SPBIBD, analogues of Theorems 4.6, 4.12 and 4.30 were obtained in [32]. In particular, [32] provided an alternative proof of Theorem 4.6. Given a SPBIBD design which is quasi-symmetric with adjacency matrices of point and block graphs given by P and A respectively, let two new graphs G_1 and G_2 be defined as follows: G_1 has adjacency matrix $\begin{bmatrix} P & N \\ N^t & A \end{bmatrix}$ and G_2 has adjacency matrix $\begin{bmatrix} 0 & j_v & 0_b \\ j_v^t & P & N \\ 0_b^t & N^t & A \end{bmatrix}$.

This is exactly the same theme as the one followed for SRGDDs earlier in this chapter but our set-up is more general. In [32], it is proved that G_1 has at most 6 and G_2 has at most 7 distinct eigenvalues. Necessary and sufficient conditions for G_1 and G_2 to be strongly regular (in terms of parameters of D) were also determined and we refer the reader to [32] for details.

The point-block regularity condition in an SPBIBD can be weakened to allow a point-block-point regularity condition. This is precisely done in the definition of a partial geometric design which was introduced by Bose, S.S. Shrikhande and Singhi [30] in order to generalize the Hall-Connor embedding theorem for $\lambda \leq 2$ (Theorem 1.17) to all values of λ.

Definition 4.40. A partial geometric design D is a 1-design (v, b, r, k) (equivalently $NJ = rJ$ and $JN = kJ$) satisfying the additional condition

$NN^tN = \theta N + tJ$ where $\theta = r + k - 1 + c - t$ for some constants t and c.

Exercise 4.41. Show that an SPBIBD is a partial geometric design. In [23], Bose, Bridges, and M.S. Shrikhande modified the proof of Theorem 4.33 (and Corollary 4.36) to obtain the following characterization.

Theorem 4.42. Let D be a 1-design with connected incidence matrix N. Then D is a partial geometric design if and only if NN^t has precisely one non-zero eigenvalue other than rk.

Proof. Let D be a partial geometric design. Put $S = NN^t$. Then $S^2 = (NN^tN)N^t = rtJ + \theta S$. We also have $JS = SJ = rkJ$. So $S^3 - \theta S^2 - r^2ktJ = 0$, i.e., $S^3 - (rk + \theta)S^2 + rk\theta S = 0$. Hence eigenvalues of NN^t are 0, rk and θ, where $\theta \neq rk$ since N is connected (and so rk is simple). Conversely, suppose $\Lambda(NN^t) = \{0, rk, \theta\}$ where rk is simple. Let $S = NN^t = A + rI$. Then A is symmetric non-negative integral matrix with zero diagonal and $\Lambda(A) = \{r(k - 1), - r, \theta - r\}$ where $r(k - 1)$ is simple. Using Theorem 2.4, $(A + rI)(A - \phi I) = mJ$ where $\phi = \theta - r$, $m = \dfrac{(d + r)(d - \phi)}{v}$, and $d = r(k - 1)$.

Now put $X = (A - \phi I)N$. Then the above equation can be used to obtain $XX^t = m(d - \phi)J$. But rank (X) = rank (XX^t) implies rank (X) = 1. Then $X = tJ$ for some t. Thus (using the Hoffman polynomial) $NN^tN = \theta N + tJ$. Hence D is a partial geometric design.

We conclude this chapter by referring to the literature. An elegant exposition of the Bose, S.S. Shrikhande and Singhi embedding theorem and related ideas can be found in [164]. Among other interesting similar structures, we mention the partial λ- geometries of Cameron and Drake [48] and semi-symmetric designs of Hughes [94].

V. STRONGLY REGULAR GRAPHS WITH STRONGLY REGULAR DECOMPOSITIONS

In the previous Chapter IV, we looked at the possibility of a strongly regular graph G containing vertex subsets which define a design or a PBIBD when the adjacency is suitably translated into incidence. A fairly general set-up was given in the notion of an SPBIBD of the last chapter. Essentially, the block graphs of interesting incidence structures are, of course, strongly regular but the combination of relations between points, blocks and point-blocks sometimes produces larger strongly regular graphs, and the procedure is often reversible. In a recent paper Haemers and Higman [73] obtained an elegant combinatorial generalization of these ideas, where a graph is partitioned into two strongly regular graphs but these graphs are not necessarily assumed to arise out of a design or a PBIBD. The present chapter is almost entirely based on the paper [73] of Haemers and Higman. Here we consider a strongly regular graph Γ_0 whose vertex set V_0 can be written as a disjoint union of two sets V_1 and V_2 such that the induced graph Γ_i on V_i, i = 1, 2 has some nice properties. These nice properties include regularity, strong regularity, being a complete graph or its complement, etc.

Definition 5.1. Let $m < n$ and let $\rho_1 \geq \rho_2 \geq \dots \geq \rho_n$ and $\sigma_1 \geq \sigma_2 \geq \dots \geq \sigma_m$ be two sequences of real numbers. Then the sequence $\sigma_1, \sigma_2, \dots, \sigma_m$ is said to interlace the sequence $\rho_1, \rho_2, \dots, \rho_n$ if the following m inequalities hold: for $i = 1, \dots, m$, $\rho_i \geq \sigma_i \geq \rho_{n-m+i}$. The interlacing is called tight if the following m equalities hold for some (fixed) k: $\rho_i = \sigma_i$ for $i = 1, \dots, k$ and $\rho_{n-m+i} = \sigma_i$ for $i = k+1, \dots, m$.

Theorem 5.2. (Haemers [68]) Let A_0 be a symmetric matrix partitioned as follows:

$$A_0 = \begin{pmatrix} A_1 & C \\ C^t & A_2 \end{pmatrix}.$$

Let B be the 2 x 2 matrix whose entries are the average row sums of the submatrices in A_0 (i.e., b_{11} is the average row sum of A_1, etc.). Then the following assertions hold:

(i) <u>Cauchy interlacing</u>: The eigenvalues of A_1 interlace the eigenvalues of A_0 and if the interlacing is tight, then $C = O$.

(ii) The eigenvalues of B interlace the eigenvalues of A_0. If the interlacing is tight, then A_1, A_2 and C have constant row and column sums. Conversely, if A_1, A_2 and C have constant row and column sums, then both the eigenvalues of B are also eigenvalues of A_0.

Exercise 5.3. In the above theorem, let A_1 be a zero matrix of order m and let α and β respectively denote the number of non-negative and non-positive eigenvalues of A_0. Show that $m \leq \min\{\alpha, \beta\}$. Hence, show that if G is a graph containing a coclique (set of mutually non-adjacent vertices) of order m, then $m \leq \min\{\alpha,\beta\}$ where α (respectively β) is the number of non-negative (non-positive) eigenvalues of G.

Lemma 5.4. (cf. [68]) Let A_i be a symmetric matrix of order v_i, $i = 0, 1, 2$ such that

$$A_0 = \begin{pmatrix} A_1 & C \\ C^t & A_2 \end{pmatrix} \text{ and } \quad A_1C + CA_2 = \alpha C + \beta J$$

for some real numbers α and β. Let A_1, A_2, C and C^t have constant row sums k_1, k_2, r and k respectively. Let the eigenvalues of A_i be ρ_{ij}, j = 1, 2, . . . , v_i, i = 0, 1, 2. Let the non-zero eigenvalues of CC^t be $\gamma_1, \gamma_2, \ldots, \gamma_m$ where m = rank of C. Then it is possible to order (i.e., relabel) the eigenvalues in such a way that the following assertions hold.

(i) $\rho_{11} = k_1$, $\rho_{21} = k_2$ and the corresponding eigenvector is the all one vector. Also, $\gamma_1 = rk$ and $k_1 + k_2 = \alpha + \beta v_1/r$.

(ii) For j = 2, . . . ,m, $\rho_{1j} + \rho_{2j} = \alpha$ and an eigenvector corresponding to ρ_{2j} is in the range of C^t, (i.e., is a multiple of C^t by a column vector) while an eigenvector corresponding to ρ_{ij} is in the range of C. Also, for j = 1, . . . , m, the eigenvalues $\rho_{0,\,2j-1}$ and $\rho_{0,\,2j}$ are the roots of $(x - \rho_{1j})(x - \rho_{2j}) = \gamma_j$.

(iii) For j = m + 1, ..., v_1, the eigenvalue ρ_{1j} has a corresponding eigenvector which is in the kernel of C^t (i.e., whose image with C^t is the zero vector), $\rho_{0j} = \rho_{1j}$, and for j = m + 1, . . . , v_1, $\rho_{0,j} = \rho_{1,j}$, for j = m + 1, . . . , v_2. Moreover, ρ_{2j} has a corresponding eigenvector which is in the kernel of C.

Proof. $A_1CC^t = \alpha CC^t + \beta rJ - CA_2C^t$ and the right hand side is a symmetric matrix of order v_1. Hence the left hand side is also a symmetric matrix, i.e., $(A_1CC^t)^t = CC^tA_1 = A_1CC^t$, i.e., A_1, CC^t, and J commute (note that A_1, C and C^t have constant row and column sums). So these matrices have a common orthonormal set of eigenvectors u_i ordered in such a way that $A_1u_j = \rho_{ij}\, u_j$ for j = 1, . . . , v_1, $(CC^t)\, u_j = \gamma_j u_j$ for j = 1, . . . , m and $(CC^t)u_j = O$ for j = m + 1, . . . , v_1, $Ju_j = O$ for $j \neq 1$, and also u_1 is the all-one vector. Therefore multiplying the given matrix equation by the "all-one" vector gives $rk_1 + rk_2 = r\alpha + \beta v_1$. This completes the proof of (i).

Transpose the given matrix equation and multiply on the right by u_j to get $A_2C^tu_j = \alpha C^tu_j + \beta Ju_j - C^tA_1u_j$ which equals $(\alpha - \rho_{1j})(C^tu_j)$ for j = 2, . . . , m (this is clear since $J_{v_1,v_1} u_j = O$ also implies $J_{v_2,v_1} u_j = O$). Hence C^tu_j is also an eigenvector of A_2 and letting the corresponding eigenvalue of A_2 be ρ_{2j} gives $\rho_{2j} = \alpha - \rho_{1j}$. Interchange of A_1 and A_2 proves the remaining part of the first sentence in (ii).

Now let

$$w_j = \begin{pmatrix} \gamma_j u_j \\ (x - \rho_{ij})C^t u_j \end{pmatrix}, j = 1, \ldots, m$$

be a vector of size $v_1 + v_2$. By actual multiplication, observe that the condition $(x - \rho_{1j})(x - \rho_{2j}) = \gamma_j$ implies that $A_0w_j = xw_j$, i.e., if x satisfies $(x - \rho_{1j})(x - \rho_{2j}) = \gamma_j$, then x is an eigenvalue of A_0. Denoting these eigenvalues by $\rho_{0,2j-1}$ and $\rho_{0,2j}$, j = 1, . . . , m gives (ii) (note that the first of these equations reads $(x - k_1)(x - k_2) = rk$).

Finally, let $w_j = \begin{pmatrix} u_j \\ O \end{pmatrix}$, $j = m + 1, \ldots, v_1$

be a vector of size $v_1 + v_2$ (where O is the zero vector of order v_2). By actual multiplication it is easily seen that $A_0w_j = \rho_{1j} w_j$. Therefore, $\rho_{0j} = \rho_{1j}$ for $j = m + 1, \ldots, v_1$. Reading of the same equation $A_0w_j = \rho_{1j} w_j$ in partitioned form gives $C^tu_j = O$ for $j = m + 1, \ldots, v_2$. The remaining part of (iii) follows by interchanging A_1 and A_2.

Definition 5.5. A strongly regular design D is a 1-design with v_1 points and v_2 blocks whose members satisfy the following additional properties: There exist graphs Γ_1 and Γ_2 (neither a complete graph nor the complement of a complete graph) with adjacency matrices A_1, and

A_2 respectively, such that the incidence matrix C of D, A_1 and A_2 satisfy all the three conditions given below:

(i) $CC^t = w_1I + y_1J + z_1A_1$ for integers w_1, y_1, z_1, and $z_1 \neq 0$.

(ii) $C^tC = w_2I + y_2J + z_2A_2$ for integers w_2, y_2, z_2, and $z_2 \neq 0$.

(iii) $CC^tC = \gamma C + \delta J$ for integers γ and δ.

Exercise 5.6. With the set-up of Definition 5.5, show that (a) C has a constant row sum $r = w_1 + y_1$ and a constant column sum $k = w_2 + y_2$ (b) $\delta = \frac{k(kr - \gamma)}{v_1}$.

Exercise 5.7. With the same set-up, show that for i = 1, 2 the graph Γ_i is strongly regular with eigenvalues $k_i = \frac{kr - y_iv_i - w_i}{z_i}$, $\rho_i = \frac{\gamma - w_i}{z_i}$ and $\sigma_i = \frac{-w_i}{z_i}$ of multiplicities 1, m - 1 and v_i - m, respectively where m = rank of C. Also, the eigenspace of the eigenvalue σ_1, (respectively σ_2) is the kernel of C (respectively C^t).

Observe that the condition (iii) in Definition 5.5 states that D is a partial geometric design of Chapter IV. Hence by Theorem 4.42, (iii) may be replaced by $\gamma_1 = rk$ and $\gamma_2 = ... = \gamma_m = \gamma$ where $\gamma \neq 0$.

Exercise 5.8. Use the above discussion and exercises to show that a strongly regular design is the same as a quasi-symmetric SPBIBD (of Definition 4.31). Clearly then Γ_1 and Γ_2 are respectively the point and block graphs of D.

At this point we make some terminology and notation clear. If G is a graph with distinct eigenvalues $\alpha_1, \alpha_2, \ldots, \alpha_s$ with multiplicities $m_1, m_2, \ldots, m_s$ respectively then the spectrum of G = spec G = $\begin{pmatrix} \alpha_1 & \alpha_2 \ldots \alpha_s \\ m_1 & m_2 \ldots m_s \end{pmatrix}$.

This chapter is concerned with strongly regular graphs Γ_0 which can be decomposed into two graphs Γ_1 and Γ_2. In most situations of interest, all of these will be strongly regular (or at least regular). If Γ_i is regular, then k_i denotes its degree, $i = 0, 1, 2$. Further, if Γ_i is strongly regular, then the other two eigenvalues of Γ_i are r_i and s_i where, by Exercise 3.10, we may assume that $r_i \geq 0 \geq s_i$. We also let f_i (respectively g_i) denote the multiplicity of r_i (respectively s_i). If Γ_1 and Γ_2 are both regular, then the decomposition is called regular and if both these graphs are strongly regular then the decomposition is called strongly regular. Also, translation of the permutation group theory terminology allows us to brand Γ_0 imprimitive if it or its complement can be expressed as a disjoint union of cliques of equal size (note that this amounts to a group divisible association scheme). A primitive Γ_0 is one which is not imprimitive and that clearly is the only interesting situation from our point of view.

Lemma 5.9. If Γ_0 is an SR graph with a regular decomposition, then $CJ = (k_0 - k_1)J$, $C^tJ = (k_0 - k_2)J$,

$$A_1^2 + CC^t = (r_0 + s_0)A_1 - r_0s_0I + (k_0 + r_0s_0)J,$$
$$A_2^2 + C^tC = (r_0 + s_0)A_2 - r_0s_0I + (k_0 + r_0s_0)J,$$
$$A_1C + CA_2 = (r_0 + s_0)C + (k_0 + r_0s_0)J.$$

(The notations in the statement of Lemma 5.9 and also essentially in this entire chapter are those of Lemma 5.4: A_i is the adjacency matrix of Γ_i and C is as given in Lemma 5.4).

Proof. Clearly, A_i has a constant row sum k_i (= the degree of Γ_i). The first equation just expresses the fact that the decomposition is regular. Since Γ_0 is strongly regular with eigenvalues k_0, r_0, and s_0 we have by Theorem 2.22, $A^2{}_0 - (r_0 + s_0)A + r_0s_0I = (k_0 + r_0s_0)J$. Reading the upper left, lower right and lower left parts in the partitioned matrix form of this equation gives the remaining equations in the statement of Lemma 5.9.

Theorem 5.10. Suppose Γ_0 is an SR graph and let Γ_1 be regular. Then

$$s_0 \leq \frac{k_1v_0 - k_0v_1}{v_0 - v_1} \leq r_0.$$

Further, the decomposition is regular if and only if equality holds on the left or right hand side. Also,

$$k_2 = \begin{cases} k_0 - k_1 + s_0, & \text{if the left hand equality holds,} \\ k_0 - k_1 + r_0, & \text{if the right hand equality holds.} \end{cases}$$

Proof. Let B denote the matrix of average row sums. Then the regularity of A_1 (i.e., Γ_1) implies that $B = [b_{ij}]$ where $b_{11} = k_1$, $b_{12} = k_0 - k_1$, $b_{21} = \dfrac{(k_0 - k_1)v_1}{v_2}$ and $b_{22} = k_0 - b_{21}$.

Clearly, the row sum of B equals k_0 and is also an eigenvalue of B. Consideration of trace gives the other eigenvalue to be $\rho = \dfrac{k_1v_0 - k_0v_1}{v_0 - v_1}$. Use of Theorem 5.2(ii) with m = 2 and i = 2 shows that $r_0 \geq \rho \geq s_0$. Also, the interlacing is tight with k = 2 (respectively k = 1) if and only if $\rho = r_0$ (respectively $\rho = s_0$) (note that the largest eigenvalue of A_0 is k_0). If A_2 is regular with degree k_2, then $b_{22} = k_2$, and since one eigenvalue of A_0 is k_0 the other must be $k_1 + k_2 - k_0$. But this eigenvalue equals ρ (which equals s_0 or r_0) and the desired conclusion is now obvious.

Exercise 5.11. With everything as in Theorem 5.10, let primes denote the corresponding parameters of the complements. Assume that the equality holds on the left (respectively right) hand side in the statement of Theorem 5.10. Then in Γ'_0 we must have

$$\frac{k_1' v_0 - k_0' v_1}{v_0 - v_1} = r_0' \text{ (respectively } s_0').$$

Theorem 5.12. Let Γ_1 be a coclique (i.e., $k_1 = 0$). Then

(i) $v_1 \leq \frac{v_0 s_0}{k_0 - s_0}$ and

(ii) if further Γ_0 is primitively strongly regular, then $v_1 \leq \min \{f_0, g_0\}$. (The first inequality is called the Hoffman coclique bound and the second inequality is called the Cvetcovic bound.)

Proof. For the first inequality, substitute $k_1 = 0$ in Theorem 5.10. Next define $A = A_0 - v_0^{-1}(k_0 - s_0) J - s_0 I$. Using the spectrum of A_0, it is easily seen that the only non-zero eigenvalue of A is $r_0 - s_0$ of multiplicity f_0. If $s_0 = 0$ then Γ_0 is imprimitive, a contradiction (use Exercise 3.10). But $A_1 = O$ and the upper left square submatrix of A is $v_0^{-1}(k_0 - s_0)J - s_0 I$ whose size is v_1 and is non-singular. So $v_1 \leq f_0$. Replacing s_0 by r_0 in this argument produces $v_1 \leq g_0$. This completes the proof of Theorem 5.12.

Suppose now that Γ_0 and Γ_1 are both SR, Γ_0 is primitive, and also assume that the decomposition is regular. Then by Theorem 5.10, $k_2 = k_0 - k_1 + \epsilon r_0 + (1 - \epsilon)s_0$ where ϵ equals 0 or 1 depending on whether the equality is met on the L.H.S. (i.e., $s_0 = \rho$) or the equality is met on the R.H.S. (i.e., $r_0 = \rho$).

Theorem 5.13. With everything as described above, one of the following holds:

(i) $s_1 > s_0$, $r_1 < r_0$, $v_1 \le \min\{f_0 + 1 - \epsilon, g_0 + \epsilon\}$, and spec Γ_2 is $\{k_2, (r_0 + s_0 - r_1)^{f_1}, (r_0 + s_0 - s_1)^{g_1}, r_0^{f_0 - v_1 + 1 - \epsilon}, s_0^{g_0 - v_1 + \epsilon}\}$.

(ii) $s_1 = s_0$, $r_1 < r_0$, $v_1 \le g_0 + \epsilon$, and spec $\Gamma_2 = \{k_2, (r_0 + s_0 - r_1)^{f_1}, r_0^{f_0 - f_1 - \epsilon}, s_0^{g_0 - v_1 + \epsilon}\}$.

(iii) $s_1 > s_0$, $r_1 = r_0$, $v_1 \le f_0 + 1 - \epsilon$, and spec Γ_2 is $\{k_2, (r_0 + s_0 - s_1)^{g_1}, r_0^{f_0 - v_1 + 1 - \epsilon}, s_0^{g_0 - g_1 - 1 + \epsilon}\}$.

Proof. Theorem 5.2 (i) shows that $s_0 \le s_1 \le r_1 \le r_0$. Using the last equation in Lemma 5.8, we have $A_1C + CA_2 = (r_0 + s_0)C^t(k_0 + r_0s_0)J$ and hence Lemma 5.4 is applicable with $\alpha = r_0 + s_0$ and $\beta = k_0 + r_0s_0$. Therefore (by Lemma 5.4 (i), (ii) and (iii)), Γ_2 has at most 5 distinct eigenvalues: k_2, $r_0 + s_0 - r_1$, $r_0 + s_0 - s_1$, r_0, and s_0. Also, if $r_1 \ne r_0$, then the eigenvalue $r_0 + s_0 - r_1$ of Γ_2 has multiplicity f_1 and if $s_1 \ne s_0$ then the eigenvalue $r_0 + s_0 - s_1$ of Γ_2 has multiplicity g_1.

We first dispose of the possibility $(r_1, s_1) = (r_0, s_0)$. Now $A_1^2 = (r_1 + s_1)A_1 - r_1s_1I + (k_1 + r_1s_1)J$ (because A_1 is SR; use Theorem 2.22). If $(r_1, s_1) = (r_0, s_0)$, then use of Lemma 5.9 shows that $CC^t = (k_0 - k_1)J$. Therefore every column of C is either a vector with all ones or all zeros. We leave it to the reader to show that this forces Γ_0 to be primitive (alternatively observe that CC^t has only one non-zero eigenvalue). Hence $r_0 = r_1$ and $s_0 = s_1$ cannot both be true. The three different cases in the statement are now obvious.

Consider (i). Then $s_1 > s_0$ and $r_1 < r_0$, and the five eigenvalues mentioned in the first paragraph of the proof are actually distinct. Notice also that r_1, r_0 are non-negative and s_1, s_0 are non-positive. The only unknowns are the multiplicities of the eigenvalues r_0 and s_0 of Γ_2. Since the trace of A_2 is zero, we have two linear equations in two unknowns (the multiplicities), the multiplicities can be completely determined, and spec Γ_2 is as given (i). The same argument also works in (ii) and (iii) since the only unknowns are the multiplicities of r_0 and s_0. This completes the proof of Theorem 5.13.

Exercise 5.14. Use Theorem 5.13 and the eigenvalue characterization of an SR graph to show that, with the hypothesis of Theorem 5.13, Γ_2 is strongly regular if and only if one of the following holds:

(i) $v_1 = f_0 + 1 - \epsilon = g_0 + \epsilon$.

(ii) $s_0 = s_1$ and $f_0 = f_1 + \epsilon$.

(iii) $s_0 = s_1$ and $v_1 = g_0 + \epsilon$.

(iv) $r_0 = r_1$ and $g_0 = g_1 + 1 - \epsilon$.

(v) $r_0 = r_1$ and $v_1 = f_0 + 1 - \epsilon$.

Theorem 5.15. Let Γ_0 be primitively SR and suppose that Γ_1 is a coclique. Then the following are equivalent

(i) Γ_2 is SR.

(ii) $v_1 = g_0 = \frac{-v_0 s_0}{k_0 - s_0}$ or $v_1 = f_0 = \frac{-v_0 r_0}{k_0 - r_0}$.

(essentially, this asserts that Γ_2 is SR if and only if both the Hoffman and Cvetcovic bounds are tight).

Proof. Assume that $v_1 = g_0 = \frac{-v_0 s_0}{k_0 - s_0}$; the other case is similar. Theorem 5.10 implies that Γ_2 is regular. Clearly $k_1 = r_1 = s_1 = 0$ since $A_1 = O$. So Theorem 5.13 (i) is applicable. Also, $\epsilon = 0$. Hence spec $\Gamma_2 = \{k_2, (r_0 + s_0)^{v_1}, r_0^{f_0 - v_1 + 1}, s_0^{g_0 - v_1}\} = \{k_2, (r_0 + s_0)^{v_1}, r_0^{f_0 - v_1 + 1}\}$. So, by the characterization Theorem 2.22, Γ_2 is SR.

Conversely, if Γ_2 is a strongly regular graph, then it is certainly regular and Theorem 5.10 implies that $v_1 = \frac{-v_0 s_0}{k_0 - s_0}$ (the other case is similar). Hence, Exercise 5.14 is applicable with $r_1 = s_1 = 0$ and $\epsilon = 0$ and spec $\Gamma_2 = \{k_2, (r_0 + s_0)^{v_1}, r_0^{f_0 - v_1 + 1}, s_0^{g_0 - v_1}\}$. But the Cvetcovic bound implies that $g_0 \geq v_1$ and $f_0 - v_1 + 1 > 0$. Since Γ_2 is SR, $g_0 = v_1$, and the proof is complete.

A conference matrix C of order n is a $(0, \pm 1)$ matrix with entries 0 only on the diagonal such that $CC^t = (n-1)I_n$. Conference matrices are similar to Hadamard matrices as the following exercise shows.

Exercise 5.16. (i) Show that the existence of a conference matrix of order n is equivalent to the existence of a conference matrix C of the same order such that C is normalized, i.e., has the first row and column (except the diagonal) entries one. (ii) Show that if n is the order of a conference matrix, then $n \equiv 2 \pmod 4$. (iii) Show that if C is a normalized symmetric conference matrix of order n and B is obtained from C by deletion of the first row and column, then the matrix $A = \frac{1}{2}(B + J - I)$ is an adjacency matrix of an SR graph with parameter set $(n', a', c', d') = (n-1, \frac{n-2}{2}, \frac{n-6}{4}, \frac{n-2}{4})$ (equivalently, A is obtained from C by deletion of the first row and column and then changing -1 to 0). (iv) Show that if A is an SR graph with parameters as in (iii), then a normalized symmetric conference matrix (with order one larger than that of A) can always be constructed.

Definition 5.17. A strongly regular graph with parameters as in (iii) is called a conference graph.

For a discussion of conference graphs, we refer the reader to [49] .

A proper strongly regular decomposition is called exceptional if $s_1 \neq s_0$ and $r_1 \neq r_0$.

Exercise 5.18. Show that if a strongly regular decomposition is exceptional, then $s_2 \neq s_0$ and $r_2 \neq r_0$.

Hint: Use Exercise 5.14 (i).

The following theorem characterizes all the exceptional SR decompositions.

Theorem 5.19. Let Γ_0 be a primitively SR graph admitting an exceptional SR decomposition. Then Γ_0 or its complement has parameters $v_0 = 4r_0^2 + 4r_0 + 2$, $k_0 = 2r_0^2 + 2r_0$, $s_0 = -r_0 - 1$, for some positive integer r_0 (we remind the reader that in this whole chapter we have let the eigenvalues represent graph parameters) and one of the following holds:

(i) Γ_1 and Γ_2 are both conference graphs with parameters $v_1 = v_2 = 2r_0^2 + 2r_0 + 1$ and D (i.e., the 1-design represented by C) is a symmetric design with parameter set $(v_1, r_0^2, \frac{r_0(r_0 - 1)}{2})$ or its complement.

(ii) $v_1 = v_2 = 2r_0^2 + 2r_0 + 1$, $k_1 = k_2 = r_0^2 + r_0$, $r_2 = \frac{k_1 - r_1}{2r_1 + 1}$, $s_1 = -r_2 - 1$, $s_2 = -r_1 - 1$, $r_1 \neq r_2$ where r_1, r_2 are both integers.

Proof. Theorem 5.13 (i) and Exercise 5.14(i) are applicable and we may assume w.l.o.g. that $\epsilon = 0$ (otherwise look at the complements) to get $v_1 = f_0 + 1 = g_0$. So $v_1 + v_2 = v_0 = f_0 + 1 + g_0$. Therefore $v_2 = g_0$, i.e., $v_2 = v_1$. It is also clear that A_1, A_2, and C have constant row sums. But C is a square matrix with row sum $k_0 - k_1$ and column sum $k_0 - k_2$. So $k_1 = k_2$. Theorem 5.12 (i) gives spec $\Gamma_2 = \{k_2, (-1-r_1)^{f_1}, (-1-s_1)^{g_1}\} = \{k_1, r_2^{g_1}, s_2^{f_1}\}$ (this is clear since r_2 is positive and s_2 is negative). Looking at the trace of A_2 gives $k_1 = \frac{v_1 - 1}{2}$. Finally, use Theorem 2.23 as given in the statement. If Γ_1 is a conference graph then $r_1 = -1 - s_1 = r_2$ and hence $r_1 = r_2$, $s_1 = s_2$, i.e., Γ_2 is also a conference graph. If not, then case (ii) is clearly obtained. This completes the proof of Theorem 5.19.

The smallest example of (i) in Theorem 5.19 is that of the Petersen graph partitioned into two 5-cycles. The smallest feasible parameter set in (ii) (where Γ_1, Γ_2 are not conference graphs) has $v_0 = 2{,}140{,}370$

according to Haemers and Higman [73]. When is a strongly regular decomposition proper but not exceptional? The answer is given in the following Theorem of Bridges and M.S. Shrikhande [32].

Theorem 5.20. Let Γ_0 be a primitively strongly regular graph with a decomposition into graphs Γ_1 and Γ_2. Let D, as before, represent the incidence structure given by C. Then this decomposition is proper and not exceptional if and only if D is a strongly regular design with point graph Γ_1 and block graph Γ_2, whose parameters satisfy

(i) $k_1 + r = k_2 + k$,

(ii) $k_1 - k \in \{\sigma_1, \rho_1 + \rho_2 - \sigma_1\}$,

(iii) $\rho_2 = \sigma_1 + z_1, \rho_1 = \sigma_2 + z_2$.

(Here k_i, ρ_i, and σ_i are eigenvalues of Γ_i.)

Proof. Let D be a strongly regular design satisfying (i), (ii) and (iii). From Definition 5.5 it is easy to see that $A_1C + CA_2$ is a linear combination of C and J (with integer coefficients) and hence Lemma 5.4 is applicable. By (i), Γ_0 is regular of degree $k_0 = k_1 + r = k_2 + k$. The first and second eigenvalues (by Lemma 5.4 (ii)) are roots of $(x - k_1)(x - k_2) = rk$ and since the first eigenvalue is k_0, the second eigenvalue must be $k_1 + k_2 - k_0$.

Observe that the eigenvalue ρ_i of Γ_1 corresponds to the range of C (by Lemma 5.4(ii)) and the other eigenvalues of Γ_0 are roots of $(x - \rho_1)(x - \rho_2) = \gamma$, where γ is the only non-zero eigenvalue of CC^t (see Exercise 5.7). The eigenvalue σ_1 corresponds to the kernel of C by Lemma 5.4(iii). Hence Γ_0 has at most five eigenvalues: $k_1 + k_2 = k_0$, the roots of $(x - \rho_1)(x - \rho_2) = \gamma$, and σ_1 and σ_2. By Exercise 5.7 and (iii) above, we have $\sigma_1 = \rho_1 - \frac{\gamma}{z_1} = \rho_2 - z_1$, so that σ_1 is easily seen to satisfy $(x - \rho_1)(x - \rho_2) = \gamma$. Similarly, σ_2 also satisfies the same equation. By looking at the sum of roots of this equation we see that if σ_1 is a root, then so is $\rho_1 + \rho_2 - \sigma_1$. Hence $k_1 + k_2 - k_0 = k_2 - r$ is a root if $k_2 - r = k_1 - k \in \{\sigma_1, \rho_1 + \rho_2 - \sigma_1\}$, which is true by (ii). All these things prove that we have a proper SR decomposition. Also, note that Γ_0 has one of σ_1, σ_2 as an eigenvalue. Hence Exercise 5.18 shows that the decomposition is not exceptional.

Conversely, assume that Γ_0 has a proper but not exceptional SR decomposition. Then $r_0 = r_1$ or $s_0 = s_1$ where w.l.o.g. we assume that $r_0 = r_1$. Then Lemma 5.4 (ii) implies that $\gamma_j = (s_0 - r_1)(s_0 - s_1) = (r_0 - s_0)(s_1 - s_0)$ for $j = 2, \ldots, m$. Hence CC^t has one non-zero eigenvalue. The discussion after Exercise 5.7 shows that D satisfies (iii) of Definition 5.5. By Lemma 5.8, CC^t is a linear combination of A_1, I, and J, while C^tC is a linear combination of A_2, I, and J. Hence D satisfies both (i) and (ii) of Definition 5.5. Therefore D is a strongly regular design. Reversing the arguments used in the first part of our proof shows that (i), (ii) and (iii) in the statement of the theorem are also satisfied.

For the proof of the following, we refer to [73].

Theorem 5.21. Let Γ_1 be a conference graph and suppose that Γ_2 is the complement of Γ_1. Further, assume that the matrices A_1 and C commute. Then Γ_0 is strongly regular with an exceptional strongly regular decomposition.

According to [73], the SR graph Γ_0 with $(v_0, k_0, r_0, s_0) = (26, 10, 2, -3)$ is the only known example of the situation in Theorem 5.21 (note: A_1 and C must commute).

Example 5.22. Consider the triangular graph T(m) (which is the line graph of K_m) defined on the vertex-set consisting of all the 2-subsets of S = {1, 2, ..., m}. Fix $x \in S$ and partition the vertices of T(m) into two sets, the first consisting of those 2-subsets which contain x and the other consisting of those 2-subsets which do not contain x. It is easily seen that the first set forms a clique Γ_1 while the second is a $T(m-1) = \Gamma_2$. We thus have an improper SR decomposition.

Example 5.23. Let q be the order of a projective plane π and suppose there is a set of q-1 mutually orthogonal latin squares of order $q^2 + q + 1$. It is well-known that the latter is equivalent to the existence of a transversal design, i.e., an SRGDD with $v' = (q + 1)(q^2 + q + 1)$, $r' = q^2 + q + 1$, $k' = q + 1$, $m' = q + 1$, $n' = q^2 + q + 1$, $\lambda_1 = 0$, and $\lambda_2 = 1$ such that every block meets every group. Let M be the incidence matrix of π and define $B_1 = I \otimes M$ where I is an identity matrix of order q + 1 and $\otimes$ the tensor product. Let B_2 be the incidence matrix of the SRGDD (written in an obvious manner). Then $B = [B_1:B_2]$ is a Steiner system with $v = (q^2 + q + 1)(q + 1)$ and block size q + 1, which is clearly quasi-symmetric. So the block graph Γ_0 is strongly regular. Let Γ_i be the subgraph on B_i, i = 1, 2. Then Γ_1 is primitively SR (actually a disjoint union of q + 1 cliques of size $q^2 + q + 1$) and Γ_2 is SR since it is block graph of a transversal design (= dual of a net). So we have a strongly regular decomposition.

Example 5.24. Let E be a quasi-symmetric 3-design. Fix a point p of E and consider the block graphs of the derived design D (say Γ_1) and the residual design R (say Γ_2) at p. Then both D and R are quasi-symmetric and the derived design D is symmetric if and only if E is a 3-design with smaller intersection number zero (use Theorem 1.29). Thus, the block graph Γ_0 of E admits a strongly regular decomposition which is improper (Γ_1 is a clique or its complement) if and only if E is the extension of a symmetric design.

We conclude this chapter by informing the reader that an infinite class of examples, based on rather involved geometric ideas, is given in [73]. The same paper contains some non-existence results and also a table of all feasible parameter sets for primitive strongly regular graphs with a strongly regular decomposition up to 300 vertices.

VI. THE WITT DESIGNS

In our earlier chapters, we have dealt somewhat at length with the questions of extensions and embeddings of (symmetric) designs, quasi-symmetric designs, and strongly regular graphs. The most elegant examples of these situations are provided by the Witt designs. Historically, statisticians were the first to make a systematic and exhaustive study of block designs, particularly from the point of view of constructions. However, block designs with regular and nice structural properties are generally obtained by making themselves available from groups. It is from this point of view that the (study of) Witt designs becomes extremely important. At present, there are at least two objects (excluding objects such as the Leech lattice) that are equivalent to Witt designs. These are the Golay codes (from coding theory) and the five Mathieu groups, all of which were the first examples of sporadic simple groups that have been completely determined now.

From the combinatorialist's point of view, a substantial portion of the research work in design theory centers around various characterizations of Witt designs by their properties. In this connection, also note that a result of Ito et al. [98, 64] and Bremner [31] implies that the only non-trivial quasi-symmetric 4-design is the Witt 4-design or its complement. It has been conjectured that the only non-trivial quasi-symmetric 3-designs (other than the Hadamard 3-designs) are the 3-designs related to the Witt designs or their complements (see Chapter IX). Therefore, the first sentence in this paragraph probably makes more sense in the context of the topics covered in this book. A second reason for the study of Witt designs is the fact that they provide examples of Steiner systems with $t \geq 4$. It was only in 1976 that the first example of a non-trivial Steiner system with $t \geq 4$ other than the Witt designs was found by Denniston [63]. Even to this date, not many examples are known and perhaps none exist with $t \geq 6$ (see [15]). A (sharply) t-transitive permutation group with a suitable set to act on and with an additional property gives rise to a t-design. Unfortunately, there is only one non-trivial sharply 4-transitive and sharply 5-transitive group as was proved by Jordan (see [15, p.195]). Though the Witt designs have a great aesthetic appeal, their example cannot possibly be emulated; these designs are an outcome of a combination of tight arithmetical conditions as was pointed out by Biggs and White [18].

To our knowledge, Mathieu groups were used by Witt [182] and Carmichael [52] in the late 1930s to construct Witt designs. Later on, in 1965, an important combinatorial elucidation of these objects was given by Lüneberg [106] in the late 1960s. Numerous semi-expository and expository articles on the Witt designs and their structural investigations can be found in the literature. This chapter is devoted to the construction and properties of Witt designs and is essentially both self-contained and purely combinatorial. Our treatment is based on the lecture notes of Sane [135], which in turn closely follows Lüneberg's [106] treatment. A similar exposition was also given by van Lint [176] and for other expositions, we refer to Beth, Jungnickel, and Lenz [15], Cameron and van Lint [49], and Hughes and Piper [95]. Starting from a projective plane of order four, we will construct the Steiner systems S(3, 6, 22), S(4, 7, 23), S(5, 8, 24), and also the smaller Witt designs. In addition to the block graphs of the quasi-symmetric designs so constructed, we will also give constructions of the Higman-Sims graph and the Hoffman-Singleton graph.

From this point on, let π denote a projective plane of order n. An α-arc A of π is a set of α points of π, no three collinear. If $\alpha = 3$

(respectively $\alpha = 4$), then A is called a triangle (respectively a quadrangle).

Proposition 6.1. Let A be an α-arc of π. Then

(i) $\alpha \leq n + 2$ with equality if and only if every line intersects A in 0 or 2 points.

(ii) If $\alpha = n + 2$, then n must be even.

(iii) If $\alpha = n + 1$ and n is even, then there is a unique point $p \notin A$ such that $A' = A \cup \{p\}$ is an $(n + 2)$-arc of π.

Proof. Call a line X a passant, tangent, or secant of A depending on whether $|X \cap A| = 0, 1$ or 2 respectively. We prove (i) and (iii) leaving (ii) to the reader as an exercise. Fix a point $a \in A$ and count the secants containing a in two ways to get $\alpha - 1 \leq n + 1$, i.e., $\alpha \leq n + 2$ with equality if and only if every line on a is a secant of A. This proves (i). For (iii), the same argument shows that a is on a unique tangent of A and hence A has $n + 1$ tangents. Let a_i be the number of points outside A that are on i tangents. Since $\alpha = n + 1$ which is odd, every point is on at least one tangent, i.e., $a_0 = 0$. Therefore $\Sigma a_i = (n^2 + n + 1) - (n + 1)$, $\Sigma i a_i = (n + 1)n$, and $\Sigma i(i - 1)a_i = (n + 1)n$, i.e., $\Sigma(i - 1)(i - (n + 1))a_i = 0$, where $1 \leq i \leq n + 1$. So $a_i = 0$ for $i \neq 1, n + 1$ and hence $a_1 + a_{n+1} = n^2$, while $a_1 + (n + 1)a_{n+1} = n^2 + n$, i.e., $a_{n+1} = 1$, thus proving (iii).

Exercise 6.2. Complete the proof of Proposition 6.1.

Following Cameron and van Lint [49], an arc A with $n + 1$ (respectively $n + 2$) points is called an oval (respectively a hyperoval). By a sub-plane π' of π we mean an incidence substructure π' (i.e., a closed subset under incidence) which is itself a projective plane (of order $m \leq n$). The following theorem is a weaker form of a classical result of Bruck [12].

Theorem 6.3. Let π' be a subplane of order $m < n$ of π. Then $n \geq m^2$ with equality if and only if every line not in π' contains a point of π' and every point not in π' is on some line of π'.

Proof. Let X be a line of π' and p a point on X such that p is not in π' (which exists since $m < n$). Then every line on p other than X contains at most one point of π'. For if not, then such a line would be a line of π' and then p would be on two lines of π', i.e., p would be in π', a contradiction. So p is on at least m^2 lines other than X, i.e., $1 + m^2 \leq n + 1$, i.e., $m^2 \leq n$, where equality occurs if and only if every line other than X containing p has a unique point of π'.

A subplane π' of order $m = \sqrt{n}$ is called a <u>Baer subplane</u> of π. Equivalently, every line of π contains a point of π' and dually.
<u>From this point on, assume that π is a projective plane of order four.</u>

Proposition 6.4. (i) Any quadrangle of π is contained in a unique hyperoval of π. (ii) Any quadrangle of π is contained in a unique Baer subplane of π. (iii) There are $3 \times 840 = 2{,}520$ quadrangles in π. (iv) There are 168 hyperovals in π. (v) There are 360 Baer subplanes in π.

Proof. (i) Let $Q = \{p, q, z, w\}$ be a quadrangle. The six lines (pq), (pz), (pw), (qz), (qw), and (zw) are distinct and contain $5 \times 6 - \binom{6}{2} + 4 = 19$ points among them, leaving us with two points a_1 and a_2 such that $Q \cup \{a_i\}$ is an oval for $i = 1, 2$. By Proposition 6.1 (iii), $Q \cup \{a_i\}$ is contained in a unique hyperoval which must be $Q \cup \{a_1, a_2\} = H$ and we are done.

(ii) Let Q and H be as in (i) and consider the line (a_1a_2) which must intersect all the lines mentioned in (i). If X denotes this line, then X contains none of p, q, z, or w. Since X is left with just $5 - 2 = 3$ points, X must contain points of intersections of (pq, zw), (pz, qw), and (pw, qz). If we denote these points by a, b, and c, then $Q \cup \{a, b, c\}$ is a Baer subplane since a, b, c are collinear.

(iii) Counting: To choose Q, we must choose p in 21 ways, q in 20 ways, z in 16 ways (avoiding the line (pq)) and w in 9 ways (avoiding the lines (pq), (pz), and (qz). Hence the number of ordered quadrangles is $21 \times 20 \times 16 \times 9$. Divide this number by 4! to get rid of the order.

(iv) Every oval contains $\binom{6}{4} = 15$ quadrangles. Now use (i) and (iii) to get the number of hyperovals to be $\frac{2520}{15} = 168$.

(v) Counting again: A Baer subplane contains 7 quadrangles. Now use (ii) and (iii).

Proposition 6.5. Let $T = \{p, q, z\}$ be a triangle. Then (i) There are three hyperovals mutually intersecting in T. (ii) There are 12 hyperovals containing p and q. (iii) There are 48 hyperovals containing p.

Proof. (i) This is very similar to Proposition 6.4. There are $21 - 12 = 9$ ways in which T can be put inside a quadrangle (avoiding the lines (pq), (pz), and (qz)). In a given hyperoval H containing T, we have $6 - 3 = 3$ quadrangles containing T. Now use Proposition 6.4 (i).

(ii) Similar to (i): There $\frac{16 \times 9}{2} = 72$ quadrangles containing both p and q while any hyperoval containing p and q has $\binom{4}{2} = 6$ quadrangles that contain both p and q. Again use Proposition 6.4(i).

Exercise 6.6. Prove Proposition 6.5 (iii).

Exercise 6.7. Let H be a (fixed) hyperoval and let $T = \{p, q, z\} \subset H$. Then the following assertions hold: (i) The number of hyperovals intersecting H in T is 2. (ii) The number of hyperovals intersecting H precisely in $\{p, q\}$ is 3. (iii) The number of hyperovals intersecting H in p alone is 12. (iv) There are 40 hyperovals intersecting H in three points, 45 intersecting H in exactly two points and 72 intersecting H in exactly one point. So there are 10 hyperovals disjoint from H.

Obvious Hint: Use Propositions 6.4 and 6.5 and inclusion-exclusion.

Now define two hyperovals H_1 and H_2 to be equivalent (written $H_1 \sim H_2$) if $|H_1 \cap H_2|$ is even. Let (H) denote the set of hyperovals equivalent to H. We would eventually like to show that $\sim$ is an equivalence relation and hence (H) is actually an equivalence class. This will be done in an indirect fashion. Clearly however, Exercise 6.7 (iv) shows that $|(H)| = 1 + 45 + 10 = 56$ for every hyperoval H.

Lemma 6.8. Let $T = \{p, q, z\}$ be a triangle and let H_i, $i = 1, 2, 3$ be the three hyperovals containing T (given by Proposition 6.5(i)). Let H be any (other) hyperoval. Then $H \sim H_i$ for some i.

Proof. Proposition 6.5 (i) shows that the set of nine points not in $H_1 \cup H_2 \cup H_3$ is actually partitioned into the lines (pq), (pz), and (qz). Let $F_i = H_i \backslash T$, $f_i = |F_i \cap H|$ for $i = 1, 2, 3$, and let $f = |H \cap T|$. Finally, let g be the number of points of H not in $H_1 \cup H_2 \cup H_3$. Since H meets every line in an even number of points, $2f + g$ is even (every point of $T \cap H$ contributes to two lines), i.e., g is even. Since $|H|$ is even, $f + f_1 +$

$f_2 + f_3$ is even and hence $(f + f_1) + (f + f_2) + (f + f_3)$ is even. Therefore, for some i, $f + f_i$ must be even.

We will now introduce some notation. P is the point-set of π and L the line-set of π. For a fixed hyperoval H, let D(H) be the incidence structure on the 21 points of π plus a new point ∞. The blocks of D(H) are 6-subsets of the following two types: (i) members of α augmented by ∞. (ii) members of L. Thus D(H) has 21 + 56 = 77 blocks. We then have:

Theorem 6.9. (i) D(H) is an S(3, 6, 22). (ii) Any two blocks of any S(3, 6, 22) intersect in 0 or 2 points. (iii) If for any two hyperovals G ~H, then (G) = (H). (iv) ~ is an equivalence relation.

Proof. Let T be a set of any three points of π. If T consists of collinear points, then T is certainly contained in a line and hence in a (unique) block of D(H). If not, then T is a triangle and is contained in three hyperovals H_i, i = 1, 2, 3 and by Lemma 6.8, $H_i \in$ (H) for some i. Hence there is some member of (H) containing T, i.e., there is a block of D(H) containing T. Thus every point triple in D(H) is contained in some block. The number of point triples is $\binom{22}{3}$ while the number of point triples covered in the 77 blocks of D(H) is $77 \times \binom{6}{3} = 77 \times 20$. The equality of these two numbers shows that every point-triple of D(H) is contained in a unique block and (i) is proved. (ii) is a consequence of Theorem 1.29 (alternatively observe that any contraction is a projective plane of order four). Let G ~ H. Then G $\in$ (H) and hence G $\in$ D(H). By (i) any block of D(H) intersects G in 0 or 2 points. So (H) is contained in (G). Since $|(G)| = |(H)| = 56$, (iii) is proved and (iv) is an immediate consequence.

Exercise 6.10. Given a block B of D(H) (or an S(3, 6, 22)), show that there are 16 blocks disjoint from it which form a symmetric (16, 6, 2) design D'. Use this to show that D(H) has no three mutually disjoint blocks.

In fact, the (unique) (16, 6, 2) design D' obtained in Exercise 6.10 can be used to show the uniqueness of the Steiner system S(3, 6, 22) roughly as follows: The 60 blocks of S(3, 6, 22) other than B and those in D' contain exactly 4 points of D' which form an oval of D'. There is a unique (16, 6, 2) design with 60 ovals and thus such a (16, 6, 2) design can be uniquely embedded in an S(3, 6, 22) (for the details we refer to [130]). Observe also that the block graph of an S(3, 6, 22) as well as its residual (in the case of D(H), it is (H)) are both strongly regular (here two blocks are adjacent if disjoint). Both the SR graphs are unique and the latter is called the Gewirtz graph. (See, e.g., [33]).

Exercise 6.11. Show that there are three equivalence classes of hyperovals each containing 56 hyperovals.

Exercise 6.12. Define a line-hyperoval of π to be a set of 6 lines, no three concurrent. Show that given a (point) hyperoval H, the six passants of H form a line hyperoval H', and given a line hyperoval H', the set of 6 points not contained on any line of H' is a hyperoval H. Thus there is a one-to-one correspondence between the (point) hyperovals and line hyperovals of H: Two (point) hyperovals have j points in common if and only if the corresponding line hyperovals have j lines in common.

Proposition 6.13. Let H_1 and H_2 be a pair of disjoint hyperovals. Consider the incidence substructure π^* whose points are the nine points of π outside $H_1 \cup H_2$ and whose blocks are the lines of π containing at least two points outside $H_1 \cup H_2$. Then π^* is an affine plane of order three. Conversely, given an affine plane π^* inside π, the set of 12 points not in π^* can be partitioned into two disjoint hyperovals $H_1 \cup H_2$ in precisely three ways (corresponding to each equivalence class of hyperovals). Therefore there are precisely 280 affine subplanes of order three in π and they form a 2-(21, 9, 48)-design with block intersection numbers 1, 3, 5.

For a proof of Proposition 6.13 refer to [137]. We now turn our attention to the construction of the Higman-Sims graph. This construction, which has been already outlined in Exercise 4.9 (it is actually an NL_2 (10) graph) essentially uses the block graph of the Steiner system S(3, 6, 22). However, we will be slightly more descriptive and follow the outline given in Biggs and White [18]. Let $\mathcal{H}_i$, i = 1, 2, 3 be the three classes of hyperovals and let $\mathcal{H} = \mathcal{H}_i$ for some i. The vertices of our graph G are ∞_1, ∞_2, the 21 points in P, the 21 lines in L, and the 56 hyperovals in $\mathcal{H}$.

Essentially, G has all the vertices of the block graph of D(H) where $H \in \mathcal{H}$ and G also contains all the points of D(H) as vertices ($\infty_2 = \infty$). In addition, G contains a new vertex ∞_1. The adjacency is defined in such a way that G contains the block graph D(H) as an induced subgraph. In addition, ∞_1 is adjacent to ∞_2 and to all the members of P (i.e., to all the points of D(H)) and the adjacency between points and blocks of D(H) is by incidence:

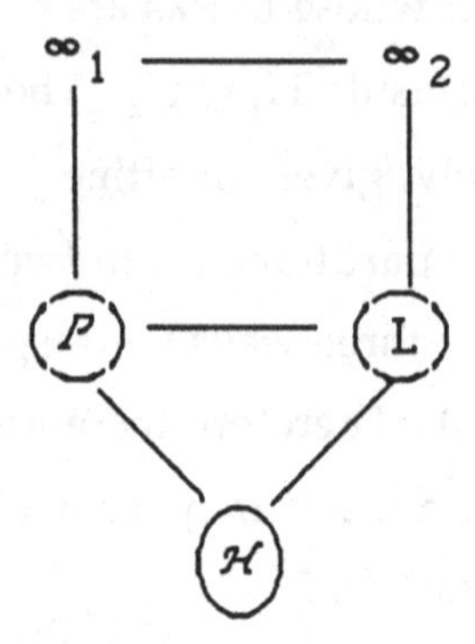

Diagram 6.14

Specifically if, $p \in P$ and $B \in L \cup \mathcal{H}$, then p is adjacent to B if p is incident to B and B_1 is adjacent to B_2 if they are disjoint. The following exercise is crucial to the proof of Theorem 6.16.

Exercise 6.15. Let B and C be two blocks of an S(3, 6, 22) such that $|B \cap C| = 2$. Then the number of blocks disjoint from both B and C is 4. Hint: Use Exercise 6.10 and Theorem 6.9 (iii).

Theorem 6.16. The Higman-Sims graph G constructed above is a strongly regular graph with the parameter set (100, 22, 0, 6).

Proof. Since $|P| = |L| = 21$, the degree of ∞_1 and ∞_2 is 22 each. Since the parameter $\lambda_1 = r$ of an S(3, 6, 22) equals 21, any vertex in P is adjacent to $1 + 21 = 22$ vertices. If $B \in L$, then B is adjacent to 16 vertices in $L \cup \mathcal{H}$ (by Exercise 6.10), 5 in P, and to ∞_2. If $B \in \mathcal{H}$, then B is adjacent to 16 vertices in $L \cup \mathcal{H}$ (by Exercise 6.10) and 6 in P. This shows that the degree of regularity = a = 22. Since the block graph of an S(3, 6, 22) is triangle-free (by Exercise 6.10), there can be no triangle in $L \cup \mathcal{H}$ and the definition of adjacency shows that if B_1 and B_2 are in $L \cup \mathcal{H}$, then they can not be commonly adjacent to a member of P. There are no triangles in P and hence c = 0 is obvious. Let $B \in L$.

Then ∞_1 and B are commonly adjacent to the 5 points of B and to ∞_2. If B $\in \mathcal{H}$, then ∞_1 and B are commonly adjacent to the 6 points of B. Essentially, the same argument holds for ∞_2 and a member of P. If B $\in \mathcal{H}$, then Exercise 6.12 shows that ∞_2 and B are commonly adjacent to 6 members of L. The parameter λ_2 of an S(3, 6, 22) equals 5. So two members of P are commonly adjacent to 6 vertices including ∞_1. Finally, if B_1 and B_2 are in $L \cup \mathcal{H}$ and are not adjacent, then $|B_1 \cap B_2| = 2$ and Exercise 6.15 shows that B_1 and B_2 are commonly adjacent to 6 vertices (including the 2 points of intersection). Therefore d = 6 and the proof of Theorem 6.16 is now complete.

The Gewirtz graph, the block graph of an S(3, 6, 22), and the Higman-Sims graph are uniquely determined by their parameters, and these three graphs seem to be the only examples of non-trivial SR graphs with $c = 0$ and $d \geq 2$.

We now turn our attention to the construction of the Steiner system S(5, 8, 24). Define two quadrangles Q and R to be equivalent if the unique hyperovals H containing Q and G containing R (by Proposition 6.4 (i)) are equivalent. We then write Q ~ R. Note that Theorem 6.9 (iv) and Proposition 6.4 (i) imply that ~ is also an equivalence relation on the set of quadrangles (we apologize for the abuse of notation) and since every hyperoval contains 15 quadrangles, every equivalence class of quadrangles has 56 x 15 = 840 quadrangles. Let $\mathcal{Q}_i$, i = 1, 2, 3 be the three equivalence classes of quadrangles where compatibility is assumed, i.e., $\mathcal{Q}_i$ is induced by $\mathcal{H}_i$. Equivalently every member of $\mathcal{Q}_i$ is contained in a member of $\mathcal{H}_i$ and conversely. We wish to extend ~ to the set of 360 Baer subplanes in a nice manner. To that end, we first prove

Lemma 6.17. Let H_1 and H_2 be two hyperovals intersecting in a set T of size 3. Write $F_i = H_i \setminus T$, i = 1, 2 and let B be any Baer subplane.

Then the following is impossible: $|B \cap T| = |B \cap F_1| = |B \cap F_2| = 2$.

Proof. Every point outside H_1 is on two passants of H_1, and any line meets H_2 in 0 or 2 points. Since H_1 has 6 passants, there are 3 = 6 - 3 passants which do not intersect $H_1 \cup H_2$. This is also clear from Exercise 6.12. Suppose $|B \cap T| = |B \cap F_1| = |B \cap F_2| = 2$. Then B has a unique point p outside $H_1 \cup H_2$. Since B is a Baer subplane, all the

three passants (common to H_1 and H_2) must contain p, i.e., p is on three passants to H_1, a contradiction.

Now define two Baer subplanes B and C to be equivalent (again written $B \sim C$) if B contains a quadrangle Q and C contains a quadrangle R such that $Q \sim R$. Since there are three equivalence classes of quadrangles, it follows that $\sim$ is an equivalence relation on the set of Baer subplanes and there are at most three equivalence classes of Baer subplanes.

Proposition 6.18. (i) Let Q and R be two quadrangles contained in a Baer subplane B. Then $Q \sim R$. (ii) For two Baer subplanes B and C, $B \sim C$ if and only if $Q \sim R$ holds for every quadrangle Q in B and every quadrangle R in C. (iii) There are precisely three equivalence classes of Baer subplanes say $\mathcal{B}_1$, $\mathcal{B}_2$, and $\mathcal{B}_3$ such that for i = 1, 2, 3, every quadrangle in $\mathcal{Q}_i$ is in some member of $\mathcal{B}_i$ and every quadrangle in every member of $\mathcal{B}_i$ is in $\mathcal{Q}_i$ (which in turn is induced by the hyperoval equivalence class $\mathcal{H}_i$). (iv) Every $\mathbf{B}_i$ contains 120 Baer subplanes.

Proof. It is an easy exercise to see that a quadrangle in a projective plane of order two is a complement of some line and hence a projective plane of order two contains 7 quadrangles forming a symmetric (7, 4, 2) design. Let Q and R be two distinct quadrangles contained in a Baer subplane B. Then there is a unique point p of B not contained in $Q \cup R$ and we may assume that (p q r), (p x_1 y_1), and (p x_2 y_2) are the three distinct lines on p in B and $Q = \{q_1, r_1, x_1, y\}$, $R = \{q_1, r_1, x_2, y_2\}$. Let H and G be the hyperovals (given by Proposition 6.4 (i)) containing Q and R respectively. Suppose H and G are not equivalent. Then $|H \cap G| \geq 2$ implies $|H \cap G| = 3$ and evidently $p \notin H \cup G$. We thus have the forbidden situation of Lemma 6.17. Clearly, the contradiction arose because of the assumption that H is not equivalent to G. So $H \sim G$ and hence $Q \sim R$. This proves (i); (ii) and (iii) are now immediate consequences. For (iv), use (i) and the fact that Q_i has 840 quadrangles.

Proposition 6.19. Let p, q, r, and s be four points such that (p q r) is a line and s is not on that line. Let x_i, i = 1, 2, 3 be the three remaining points on the line (rs). Then there are precisely three Baer subplanes B_i, i = 1, 2, 3 containing p, q, r, and s that can be ordered in such a way that x_j is in B_i if and only if j = i. Also the B_i's are mutually inequivalent.

Proof. Let Q_i denote the quadrangle $\{p, q, s, x_i\}$ and let B_i be the unique Baer subplane containing Q_i (given by Proposition 6.4 (ii)). Then B_i contains r since r is the point of intersection of (pq) and (sx_i). On the other hand, any B containing p, q, r, and s must contain one more point of the line (rs), say x_i, and then $B = B_i$. Finally, observe that Q_1, Q_2, and Q_3 mutually intersect in three points and hence must be inequivalent. So the B_i's are inequivalent and the proof is complete.

Theorem 6.20. Let D be the incidence structure whose point-set is $P \cup \{\infty_1, \infty_2\}$ and whose blocks are 7-subset (of the 23-set) of the following four types: (i) $X \cup \{\infty_1, \infty_2\}$, where X is a line of π, (ii) $X \cup \{\infty_1\}$, where X is a hyperoval in $\mathcal{H}_1$, (iii) $X \cup \{\infty_2\}$, where X is a hyperoval in $\mathcal{H}_2$, (iv) a Baer subplane $B \in \boldsymbol{B}_3$ (thus D has 21 + 56 + 56 + 120 = 253 blocks). Then D is an S(4, 7, 23).

Proof. This follows by using Proposition 6.19.

Lemma 6.21. Let C be a set of 7 points of π with the property that C intersects every line of π in 1 or 3 points or C does not intersect any line of π in 2 points. Then C determines the point-set of a Baer subplane of π.

Proof. Let a_i denote the number of lines intersecting C in i points. Then a simple counting produces Σ (i - 1)(i - 3)a_i = 0. If $a_2 = 0$, then each summand is non-negative and hence $a_i = 0$ for $i \neq 1, 3$. Clearly then, $a_1 = 14$ and $a_3 = 7$. It suffices to show that any two lines intersecting C in three points intersect each other in C. If not, then we get collinear points r_1, r_2, r_3, and s_1, s_2, s_3 such that $(r_1\ r_2\ r_3)$ and $(s_1\ s_2\ s_3)$ intersect in u and $r_i, s_j \in C$, i, j= 1, 2, 3. Let p be the remaining seventh point of C. The lines $(r_i s_j)$ are nine distinct lines all of which must contain p by the given condition. This contradiction proves Lemma 6.21.

Proposition 6.22. (i) Any Steiner system S(4, 7, 23) is essentially obtained from a projective plane π of order four using hyperovals in $\mathcal{H}_i$, $\mathcal{H}_j$ and Baer subplanes in $\boldsymbol{B}_m$ in the manner described in Theorem 6.20 (where (i, j, m) is a permutation of (1, 2, 3)). (ii) Any Steiner system S(4, 7, 23) is quasi-symmetric with the block intersection numbers 1 and 3.

Proof. Notice that contraction of D (an S(4, 7, 23)) at any point produces an S(3, 6, 22) which is quasi-symmetric by Theorem 6.9 (ii), and contraction at two points produces an S(2, 5, 21), i.e., a projective plane (whose lines intersect in one point). Hence it is clear that no two blocks of D can intersect in exactly two points. Let ∞_1 and ∞_2 be two distinguished points. A simple counting shows that there are 21 blocks containing ∞_1 and ∞_2 and 56 containing ∞_i but not ∞_j, i, j = 1, 2, and hence 120 blocks not containing ∞_1 and ∞_2. Contraction of D at ∞_2 shows that we have 56 subsets each of size 6 which intersect all the lines (using Theorem 6.9 (ii)) and among themselves in 0 or 2 points. Theorem 6.9 and Exercise 6.11 shows that the blocks containing ∞_2 but not ∞_1 must come from the other hyperoval class, say $\mathcal{H}_2$. Let $\boldsymbol{B}$ be the set of remaining 120 blocks. Then every member B of $\boldsymbol{B}$ is a 7-subset of points and by our initial assertion B does not intersect any line of π in two points. Hence by Lemma 6.21, B is a Baer subplane. Thus $\boldsymbol{B}$ is a set of 120 Baer subplanes.

Let the equivalence classes of quadrangles covered by the members of $\mathcal{H}_1$ and $\mathcal{H}_2$ be Q_1 and Q_2. Since every 4-tuple is covered precisely once in D, it follows that all the quadrangles in Q_3 are covered by Baer subplanes in $\boldsymbol{B}$, and since the members of $\boldsymbol{B}$ cover $120 \times 7 = 840 = |Q_3|$ quadrangles, $\boldsymbol{B} = \boldsymbol{B}_3$. We have thus proved (i).

Since no two blocks of D intersect in 4 or more points, the proof of (ii) is complete if we can show that no two blocks are disjoint. Let X and Y be any two blocks of D. Then by relabelling, we may assume that $\infty_1, \infty_2 \in X$ and $\infty_1, \infty_2 \notin Y$. If $X' = X \setminus \{\infty_1, \infty_2\}$, then X' is a line of π and Y a Baer subplane. So $|X \cap Y| = |X' \cap Y| = 1$ or 3.

Exercise 6.23. Use a variance argument to prove Proposition 6.22.

Presuming the uniqueness of a projective plane π of order 4 (see Cameron and van Lint [49]), our discussion so far shows that there are at most three S(3, 6, 22) and S(4, 7, 23). In fact, these Steiner systems are unique and this can be shown by using an automorphism that permutes the equivalence classes of hyperovals. Also the block graph of an S(4, 7, 23) is uniquely determined by its parameters. If we restrict ourselves to the incidence structure D' on the Baer subplanes in $\boldsymbol{B}_3$, then it is easily seen that D' is a 2-(21, 7, 12) design which is quasi-symmetric by Proposition 6.22. Hence, we get an SR graph on 120 vertices.

We now go on to construct an S(5, 8, 24). In fact, since such an object D must be an extension of π (a projective plane of order) 4 by three points $\infty_1, \infty_2, \infty_3$, the following discussion will show that D is unique and will also construct it.

Exercise 6.24. Show that any two blocks of D (i.e., an S(5, 8, 24)) intersect in 0, 2 or 4 points.

Hint: Use Proposition 6.22 (ii).

Exercise 6.25. Show that, if F is a Steiner system S(3, 6, 22) and ∞ a distinguished point of F, then the 56 blocks not containing ∞ form a class of hyperovals of the projective plane obtained by contraction at ∞.

Now use Theorem 1.5 to compute the other parameters of D (= an S(5, 8, 24)): $\lambda_4 = 5, \lambda_3 = 21, \lambda_2 = 77, \lambda_1 = r = 253$ and $\lambda_0 = b = 759$. Hence there are 77 - 21 = 56 blocks containing ∞_2 and ∞_3 but not containing ∞_1. Truncate these 56 blocks (by removing ∞_2 and ∞_3) and denote this set of 56 6-subsets of the point-set of π by S. Contraction of D at ∞_2, ∞_3 shows (using Exercise 6.25 and Exercise 6.24) that S is an equivalence class of hyperovals, say $\mathcal{H}_1$. Similarly the blocks containing ∞_1 and ∞_3 (but not ∞_2) correspond to $\mathcal{H}_2$, and the blocks containing ∞_1 and ∞_2 (but not ∞_3) correspond to $\mathcal{H}_3$. Therefore we have already determined 21 + 56 + 56 + 56 = 189 blocks of D.

Now contract D at ∞_3 and use Proposition 6.22 to conclude that the 120 blocks that contain ∞_3 alone actually come from $\mathbf{B}_3$. Similarly the 120 blocks that contain ∞_1 alone (respectively ∞_2 alone) come from $\mathbf{B}_1$ (respectively $\mathbf{B}_2$). We have thus completely determined the blocks containing ∞_1, ∞_2, or ∞_3 uniquely (as 8-subsets of the point-set of D).

It is easily seen that we are now left with 759 - (21 + 56 + 56 + 56 +120 + 120 + 120) = 210 blocks that are subsets of the point-set of π which are yet to be determined.

Let X and Y be two lines of π that intersect in p. Write $X' = X \setminus \{p\}$ and $Y' = Y \setminus \{p\}$. Suppose $s \in Y'$. Then the 5-tuple $X' \cup \{s\}$ is not covered by any hyperoval or Baer subplane since it contains 4 collinear points. Let A be the unique block containing this 5-tuple (clearly A must be among the remaining 210 blocks) and suppose $u \in A$ such that $u \neq s$ and $u \notin X'$. Then the line (us) must meet X in some point, say q. If $q \in X'$, then A covers the 5-tuple $\{u, s, q, r_1, r_2\}$ which is contained in some Baer subplane by Proposition 6.19 (choose r_1, r_2 on X' and $r_1, r_2 \neq q$). This contradiction shows that $q \notin X'$ and hence $q = p$. Therefore u is on the line (ps) = Y. It is now clear that A is contained in $X' \cup Y'$ and $|A| = |X' \cup Y'| = 8$ forces $A = X' \cup Y'$. We obtain $\binom{21}{2} = 210$ such blocks (for every choice of X and Y). But we are just left with 210 blocks and hence the remaining 210 blocks are determined as 8-subsets of the point-set of π. This completes the proof of the following result.

Theorem 6.26. Up to isomorphism there is a unique S(5, 8, 24) (which is constructed in the manner explained above).

Exercise 6.27. Show that the complement of an S(5, 8, 24) is a 5-(24, 16, 78) design.

As a byproduct, we now construct the Steiner system S(5, 6, 12).

Lemma 6.28. Let D be the Steiner system S(5, 8, 24). Then the following assertions hold:

(i) For any two points p and q and two blocks X and Y with $X \cap Y = \{p, q\}$ there is no block Z for which $X \cap Z = Y \cap Z = \{p, q\}$.

(ii) Let X be any block and let F be the incidence structure on the set of 30 blocks disjoint from X. Then F is a 3-(16, 8, 3) design. Hence every two blocks of F intersect in 0 or 4 points.

(iii) Let X,Y be a pair of disjoint blocks. Then the set of 8 points not contained in $X \cup Y$ is a block.

(iv) Let A and B be two blocks intersecting in two points. Then there is no block X disjoint from both A and B.

Proof. (i) follows by contraction at p and q and the fact that an S(3, 6, 22) has no three mutually disjoint blocks (Exercise 6.10). Consider (ii) and let T be a fixed point-triple outside X. Let a_i be the number of blocks containing T that intersect X in i points. Then Exercise 6.24 gives $a_0 + a_2 + a_4 = 21$, $2a_2 + 4a_4 = 8 \times 5 = 40$, and $a_2 + 6a_4 = \binom{8}{2} = 28$, whence $a_0 = 3$ is obtained. So F is a 3-(16, 8, 3) design which is quasi-symmetric by Theorem 1.28 and has intersection numbers 0 and 4 (actually F is a Hadamard 3-design). (iii) is then an immediate consequence (of the fact that F is resolvable). For (iv), suppose X is disjoint from both A and B. Then A and B are blocks of the 3-design F in (ii) and hence A and B must intersect in 0 or 4 points, a contradiction.

Theorem 6.29. Let A and B be two blocks of D (= S(5, 8, 24)) such that $|A \cap B| = 2$. Let R_1 be the set $(A\backslash B) \cup (B\backslash A)$ and let R_2 be the complement of R_1 (in the point-set of D). Suppose E_1 (respectively E_2) is the incidence structure whose blocks are all those blocks of D that contain at least 5 points of R_1 (respectively R_2). Then both E_1 and E_2 are Steiner systems S(5, 6, 12).

Proof. For a block X, let X_1 (respectively X_2) denote $X \cap R_1$ (respectively $X \cap R_2$). If $|X_1|$ is odd, then $|X \cap (A\backslash B)|$ and $|X \cap (B\backslash A)|$ have different parity and by Exercise 6.24, $|X \cap A \cap B|$ is both even and odd which is impossible. So $|X_1|$ and $|X_2|$ are both even. If $|X_1| = 0$, then $X = X_2$. In that case X is either disjoint from both A and B and we contradict Lemma 6.28 (iii) or X contains one or both the points of $A \cap B$ and we contradict Lemma 6.28 (i). Hence $|X_1| \neq 0$. Let $|X_1| = 8$. Then $X = X_1$. Let F be the 3-design (consisting of the blocks disjoint from X) given in Lemma 6.28 (ii). Suppose $A \cap B = \{p, q\}$. Then there are 7 (= λ_2 of F) blocks containing p and q and disjoint from X. Every such block must contain some point of R_1 (by Lemma 6.28 (i)). Hence each of these 7 blocks must contain at least two (by Exercise 6.24) points of R_1. But λ_3 (of F) equals 3 and we have only 4 points left in R_1 (other than those in $X = X_1$). So $7 \times 2 \leq 3 \times 4$ which is a contradiction. Therefore $|X_1| \neq 8$. Hence $|X_1| = 2, 4$ or 6. It is now clear that every block of E_1 and E_2 has size 6 and the proof of Theorem 6.29 is complete.

Point-sets R_1 and R_2 (which are complements of each other) are both induced by the pair (A,B) of blocks with $|A \cap B| = 2$. Such point-subsets are called dodecads. How many dodecads are there? We will eventually answer this question after noting

Exercise 6.30. Any S(5, 6, 12) is resolvable, and given any block its complement is also a block. Thus we have 66 pairs of disjoint blocks.

Now we turn our attention back to E_1 (or E_2) constructed in Theorem 6.29. Let X_1 and Y_1 be a pair of disjoint blocks (given by Exercise 6.30) of E_1, and with the notation of the proof of Theorem 6.29, let X (respectively Y) induce X_1 (respectively Y_1). Then either $|X \cap Y|$ equals 2 or X and Y are disjoint. In the latter case the unique block Z (given by Lemma 6.28 (iii)) is contained in R_2, i.e., E_2 has a block of size 8, which contradicts Theorem 6.29. Hence $|X \cap Y| = 2$ and both the points of intersection of X and Y are in R_2. If X_1' and Y_1' is some other disjoint block-pair of E_1, then the same argument shows that $|X' \cap Y'| = 2$ and both the points of intersection of X' and Y' are in R_2. Since $X_1 \cup Y_1 = R_1$, w.l.o.g. X_1' intersects X_1 in at least three points. If $X' \cap Y' = X \cap Y$, then X' intersects X in 3 points of R_1 and 2 points of R_2, a contradiction unless X = X'. So X' = X. But then Y' = Y. This argument shows that distinct block-pairs in E_1 give rise to distinct point-pairs in R_2. But E_1 has 66 such block-pairs and R_2 has $\binom{12}{2} = 66$ point-pairs. This proves that E_1 (i.e., R_1) is also induced by any point-pair of R_2. It now follows that R_1 is induced by any block-pair (X_1,Y_1) where $X_1 \cap Y_1 = \emptyset$ and X_1 contains a given 5-tuple of points of R_1. We have thus proved the following.

Theorem 6.31. The set of all 2576 dodecads in an S(5, 8, 24) forms a 5-(24, 12, 48) design.

Proof. Indeed, our observation shows that it is sufficient to catch hold of a block X containing a given 5-tuple T, if we have to find $\lambda = \lambda_5$. Let $X = T \cup \{p_1, p_2, p_3\}$. Then for any $i \neq j$ and $i, j = 1, 2, 3$ let Y be a block intersecting X in p_i and p_j alone. We know that $(X \backslash Y) \cup (Y \backslash X)$ is a dodecad (and certainly contains T). Using Exercise 6.24, Y can be chosen in $(77 - 1) - \binom{6}{2}(\lambda_4 - 1) = 76 - 60 = 16$ ways and $\{p_i, p_j\}$ in three ways. Hence $\lambda_5 = 48$ as asserted. It now easily follows that there are exactly 2576 dodecads.

We now give a construction of the Hoffman-Singleton graph, i.e., an SR graph H with parameter set (50, 7, 0, 1) from the Higman-Sims graph G. Recall diagram 6.14. This construction seems to have been first observed by Sims [165] and was made somewhat more explicit by Brouwer and van Lint [33]. We invite the reader to prove the following exercise; most of the parts are merely to be found in our earlier discussions.

Exercise 6.32. Let D be the S(3, 6, 22) constructed using π and the class $\mathcal{H}_1$ of hyperovals as in Theorem 6.9. Let B be a Baer subplane in the equivalence class $\mathbf{B_3}$ (Theorem 6.20). Then prove the following assertions.

(i) No three blocks of D are mutually disjoint.

(ii) Given a block X of D, the set of blocks disjoint from it forms a symmetric (16, 6, 2) design.

(iii) Let X and Y be two blocks of D such that $|X \cap Y| = 2$. Then for every point $p \notin X \cup Y$, the number of blocks on p that are disjoint from $X \cup Y$ is 2. Hence there are 4 blocks disjoint from both X and Y.

(iv) For $p \in B$, the number of blocks of D that meet B in p alone is 6; of these two are lines and 4 are hyperovals.

(v) Given a block X of D such that $|X \cap B| = 1$ and a point $p \in B$ with $p \notin X$, there is a unique block Y of D such that $p \in Y$ and Y is disjoint from X.

Theorem 6.33. Consider Diagram 6.14. Let G be the Higman-Sims graph, and (with the notations and terminology of Exercise 6.32) let H be the induced subgraph of G on the following 50 vertices: ∞_1, 7 points in P that belong to B and 42 blocks, of D each intersecting B in a single point (use Exercise 6.32 (iv)). Then H is an SR graph with parameter set (50, 7, 0, 1).

Proof. The degree of regularity for a point and a block is respectively shown by Exercise 6.32 (iv) and (v). Since H is a subgraph of G, c must be 0. If X is a block (and a vertex in H), then the definition of H shows that ∞_1 and X are commonly adjacent to a unique point of B. Again by definition, two points are not simultaneously adjacent to any block (for such a block must intersect B in two points) and are simultaneously adjacent to ∞_1 alone. If p and X are non-adjacent, where p is a point and X a block, then Exercise 6.32(v) shows the existence of a unique Y commonly adjacent to both p and X. Finally let X and Z be two blocks that are not adjacent. Then $|X \cap Z| = 2$ and we need to make two cases depending on whether $|X \cap Z \cap B| = 1$ or 0. In the first case, let $p \in X \cap Z \cap B$. Then p is adjacent to X and Z; suppose a_i is the number of blocks disjoint from both X and Z and intersecting B in i points. Then, by Exercise 6.32 (iii), $a_1 + a_3 = 4$ and $a_1 + 3a_3 = 6 \times 2 = 12$, i.e., $a_1 = 0$ and there is no block (in H) commonly adjacent to both X and Z. In the second case, with the same meaning given to a_i's, we obtain $a_1 + a_3 = 4$ and $a_1 + 3a_3 = 5 \times 2 = 10$, i.e., $a_1 = 1$ and there is a unique block (in H) which is commonly adjacent to X and Y. This shows that d = 1 and our proof is complete.

Exercise 6.34. Show that, if H' is the induced subgraph on the remaining 50 vertices of G, then H' is an SR graph with the same parameters.

In fact, Brouwer and van Lint [33] assert that the above procedure is the only way in which the Higman-Sims graph can be partitioned into two disjoint Hoffman-Singleton graphs. Since B can be chosen in 120 + 120 = 240 ways, this means that there are 240 disjoint pairs of Hoffman-Singleton graphs in the Higman-Sims graph. We close the chapter by informing the reader that there are some other strongly regular graphs and incidence structures that can be obtained from the Witt designs, and the most noteworthy among these is the McLaughlin graph, which can be obtained from the block graph of the S(4, 7, 23) after a certain "switching" procedure (see [33] for the details).

VII. EXTENSIONS OF SYMMETRIC DESIGNS

An important class of quasi-symmetric designs is the class of symmetric designs characterized among the former larger class by the property of having only one block intersection number. Though the symmetric designs by themselves are improper quasi-symmetric designs, as we already saw in Theorem 1.29, the extendable symmetric designs open up many possibilities for the parameter sets of proper quasi-symmetric designs. In fact, the classification theorem of Cameron (Theorem 1.29) has given rise to a considerable activity in the area of quasi-symmetric 2 and 3-designs. We will choose to postpone these topics to later chapters and concentrate here on the structure of those 3-designs that can be obtained as extensions of symmetric designs. In doing so, we will also consider some other quasi-symmetric designs (such as a residual design) associated with the extension process.

Recalling Cameron's Theorem, observe that it classifies extendable symmetric designs into four sets: an infinite set of symmetric designs the first object of which is a projective plane of order four, an infinite set of all the Hadamard 2-designs, a projective plane of order ten and a symmetric (495, 39, 3)-design. The existence of a Hadamard 2-design is equivalent to the existence of the Hadamard matrix of corresponding order. Nothing is known about a (495, 39, 3)-design. In this chapter, we first consider the extension question of a projective plane of order ten. We then consider an important class of symmetric (v, k, λ) - designs with $v = \lambda^2(\lambda + 2)$, $k = \lambda^2 + \lambda$ which is important both from the point of view of the existence question of symmetric designs and the extension problem of the infinite class of symmetric designs mentioned above. We then study the extension question of symmetric designs in that infinite class and as an application give a recent result of Bagchi [8] which shows that a symmetric (56, 11, 2)-design has no extension.

Exercise 7.1. Let D be an extension of a projective plane π of order n where $\pi = D_\infty$ and let R denote the residual design (w.r.t. ∞). Then prove the following: (i) R is a 2 - ($n^2 + n + 1$, n + 2, n) design with block intersection numbers 0 and 2. (ii) D is a quasi-symmetric 3-design with block intersection numbers 0 and 2. (iii) The blocks of R are the hyperovals of π.

Theorem 7.2. [133] Let R be a quasi-symmetric 2-($n^2 + n + 1$, n + 2, n) design with $n \geq 2$. Then n = 2, 4, or 10, and there exists a unique projective plane π of order n and (consequently) a 3-design D such that $D_\infty = \pi$ and R is the residual at ∞. Consequently R exists if and only if there is a projective plane of order n with an extension.

Proof. Clearly the other parameters of R are $r = n^2$ and $b = \dfrac{n^2 (n^2 + n + 1)}{n + 2}$. So n = 2, 4, or 10. Let (p, B) be a flag and suppose a_x (respectively a_y) is the number of blocks containing p that intersect B in x (respectively y) points. First note that x = y if and only if R is symmetric which is true if and only if n = 2 (and then x = y = 2). So let n = 4 or 10. In that case R is not symmetric and $a_x + a_y = n^2 - 1$, $(x - 1)\, a_x + (y - 1)\, a_y = (n + 1)(n - 1)$, i.e., $(x - 2)\, a_x + (y - 2)\, a_y = 0$. This shows that if y = 2, then $a_x = 0$ and since R is not symmetric therefore x = 0. Suppose $y \neq 2$. If x = 1, then by Corollary 3.9, y - 1 divides k - 1 = n + 1 = 5 or 11. Since $y \neq 2$, y - 1 = k - 1, i.e., y = k and this implies $x = \lambda = n = 1$, which is a contradiction. Also $x \neq 1$ implies x = 0 and then y = 2. Therefore (x, y) = (0, 2).

For a pair (α, β) of points of R define $\mathcal{L}_{\alpha,\beta} = \{\gamma\colon \gamma = \alpha \text{ or } \gamma = \beta \text{ or } \gamma \text{ is not contained in any block containing both } \alpha \text{ and } \beta\}$. Since y = 2, we have $|\mathcal{L}_{\alpha\beta}| = v - \lambda(k - 2) = (n^2 + n + 1) - nn = n + 1$. Suppose $\gamma \in \mathcal{L}_{\alpha\beta}$, γ

$\neq \alpha, \beta$. Let B_i (respectively C_j), $i, j = 1, 2, ..., n$ denote the blocks containing the pair (α, β) (respectively (α, γ)). Since $\gamma \in \mathcal{l}_{\alpha\beta}$, $\beta \in \mathcal{l}_{\alpha\gamma}$, therefore B_i's and C_j's are all distinct and B_i's do not contain γ while C_j's do not contain β. Hence every B_i intersects every C_j in one more point other than α and since C_j's mutually intersect in $\{\alpha, \gamma\}$, we obtain n distinct points of intersection. So $B_i \setminus \{\alpha, \beta\}$ is contained in $\cup (C_j \setminus \{\alpha, \gamma\})$ and $\cup B_i \setminus \{\alpha, \beta\} = \cup C_j \setminus \{\alpha, \gamma\}$. Hence any point of $\mathcal{l}_{\alpha\beta}$ other than α, β and γ is not contained in $\cup (C_j \setminus \{\alpha, \gamma\})$. But any point outside $\cup (C_j \setminus \{\alpha, \gamma\})$ is in $\mathcal{l}_{\alpha\gamma}$. So $\mathcal{l}_{\alpha\beta}$ is contained in $\mathcal{l}_{\alpha\gamma}$ and cardinality shows that $\mathcal{l}_{\alpha\beta} = \mathcal{l}_{\alpha\gamma}$. Now, if δ is some other point of $\mathcal{l}_{\alpha\beta}$, then $\mathcal{l}_{\alpha\gamma} = \mathcal{l}_{\alpha\beta}$ and $\mathcal{l}_{\alpha\delta} = \mathcal{l}_{\alpha\beta}$. Hence $\mathcal{l}_{\alpha\gamma} = \mathcal{l}_{\alpha\delta}$, i.e., $\delta \in \mathcal{l}_{\alpha\gamma}$ and so $\mathcal{l}_{\alpha\gamma} = \mathcal{l}_{\delta\gamma}$. Therefore $\mathcal{l}_{\alpha\beta} = \mathcal{l}_{\alpha\gamma} = \mathcal{l}_{\gamma\delta}$. <u>Hence for any point-pair (γ, δ) contained in $\mathcal{l}_{\alpha\beta}$, $\mathcal{l}_{\alpha\beta} = \mathcal{l}_{\gamma\delta}$.</u>

Let π be the incidence structure whose blocks are $\mathcal{l}_{\alpha\beta}$. Then from what we just proved, π is a projective plane of order n. Also any point-triple which is not contained in a block of R is contained in a unique block of π (this also shows that π is unique). Therefore augmenting every block of π by ∞ obtains a 3-design D with R as the residual and completes the proof of Theorem 7.2.

An important result (whose known proof depends on a computer search) of Lam et al. [101] proves that no projective plane of order ten can have any hyperoval and hence in particular there are no extendable projective planes of order ten. This at once obtains the following.

Corollary 7.3. A quasi-symmetric 2 - (111, 12, 10) design (with (x, y) = (0, 2)) does not exist.

(Lam et al. [102] have recently apparently proved that no projective plane of order ten exists).

We now turn to symmetric designs where the projective planes give an infinite class of known examples with $\lambda = 1$. For a parameter set (v, k, λ) to be feasible it must satisfy the necessary Bruck-Ryser-Chowla conditions. (Theorems 1.15, 1.16) Moreover, these conditions are probably not as strong in view of the fact that the present knowledge about 'existing' symmetric designs seems to be limited to the following:

Fact 7.4. Let (v, k, λ) be a parameter set with $\lambda(v - 1) = k(k - 1)$ and $\lambda \geq 2$. Suppose a symmetric (v, k, λ) - design is 'known' to exist. Then with precisely four exceptions, in all the known configurations we have $k \leq \lambda^2 + \lambda$. The exceptions are given by: (37, 9, 2), (56, 11, 2), (79, 13, 2) and (71, 15, 3).

We caution the reader that the above fact does not state $k \leq \lambda^2 + \lambda$ for $\lambda \geq 2$ (with four exceptions) as a theorem. It merely tells us that this is the case in all the examples that have been 'constructed so far.' This prompts us to make the following:

Definition 7.5. A symmetric (v, k, λ) design with $\lambda \geq 2$ and $k = \lambda^2 + \lambda$ is called a threshold symmetric design. We denote a threshold symmetric design (with the above parameters) by F_λ.

Fortunately, threshold symmetric designs exist for all the values of λ for which a projective plane of order λ exists. If λ is a prime power, this fact has been reproved by several people [136]. In fact, one can actually construct (the stronger) a (v, k, λ)-graph, i.e., an SR graph with $\lambda^2(\lambda + 2)$ vertices, $a = \lambda^2 + \lambda$ and $c = d = \lambda$ (see [1]). If λ is an order of a projective plane, then this seems to have been first observed by Lenz and Jungnickel [104]. The construction which uses an auxiliary set of matrices is described below.

Theorem 7.6. Let A be an affine plane of order q. Suppose $l_1, l_2, \ldots, l_{q+1}$ are the q + 1 parallel classes of lines of A. Label the points of A by 1, 2, ..., q^2. Form the q + 1 partition matrices $A_\alpha = [a_{ij}^\alpha]$, $\alpha = 1, 2, \ldots, q+1$, each of order q^2 by letting $a_{ij}^\alpha = 1$ if the points i and j are commonly contained on a line in the α-th parallel class and write $a_{ij}^\alpha = 0$ otherwise. Now let S be a latin square of order q + 2, $S = [s_{\alpha\beta}]$, $\alpha, \beta = 0, 1, 2, \ldots, q+1$, where the entries in S are the integers 0, 1, 2, ..., q + 1. That is, S is a matrix in which every row and column is a permutation of $\{0, 1, 2, \ldots, q+1\}$. Now define a (block) matrix N of order $q^2(q+2)$ by letting $N = [N_{\alpha\beta}]$ where $N_{\alpha\beta} = A_\gamma$ if $s_{\alpha\beta} = \gamma$. Here A_0 is the zero matrix of order q^2 and $\alpha, \beta, \gamma = 0, 1, 2, \ldots, q+1$. Then N represents an incidence matrix of an F_q.

Exercise 7.7. Prove Theorem 7.6. Proceed through the following steps: (i) A_α is symmetric with 1 on the diagonal for all $\alpha \neq 0$. (ii) For $i \neq j$, there is a unique α for which $a_{ij}^\alpha = 1$. (iii) If $\alpha \neq 0$, then the inner product of any two rows of A_α is 0 or q. (iv) For $\alpha \neq 0$, $A_\alpha A^t_\alpha = q^2 I + qJ$. (v) For $\alpha, \beta \neq 0$ and $\alpha \neq \beta$, $A_\alpha A^t_\beta = J$. So, $NN^t = q^2 I + qJ$.

By choosing S to be a symmetric latin square one can construct an F_q in Theorem 7.6 such that the constructed symmetric design has a polarity. In fact, a slight modification also yields a (v, k, λ) - graph. If we restrict the incidence matrix N in Theorem 7.6 to any of the $q^2 \times (q^2(q+2))$ submatrices corresponding to a row of S, then every non-zero column is seen to be repeated q times and this gives rise to q (q + 1) distinct columns which form lines of the affine plane *A* we started with. Prompted by this observation , the following definition was made by Sane [136].

Definition 7.8. A threshold symmetric design $D = F_t$ is called quasi-affine if its point-set can be partitioned into t + 2 subsets $P_1, P_2, \ldots, P_{t+2}$ such that the following holds for every $i = 1, 2, \ldots, t + 2$: The incidence structure $A(P_i)$ induced by the non-empty intersections of blocks of D with P_i is an affine plane. Here, we identify two (truncated) blocks in $A(P_i)$ if (and only if) they have identical intersection with P_i.

We recall the following well known concept [15]: A line (p, q) of any symmetric design D is the intersection of all the λ blocks containing both p and q. Clearly the size of any line is at most λ and at least two. Hence, if $|(p, q)| = \lambda$ (respectively 2), then (p, q) is called a full (respectively a trivial) line. Notice that the construction of Theorem 7.6 produces an F_t with a large number of full lines (use Exercise 7.7 (iii)). Quasi-affine designs were investigated in [136] and the following two theorems list some of the findings.

Theorem 7.9. Let $D = F_t$ be a quasi-affine symmetric design. Then for $t \neq 2$, the point-set of D can be partitioned into the subsets P_i's in a unique manner and every $A(P_i)$ is an an affine plane of order t. Every line of every $A(P_i)$ is induced by t blocks of D and thus D has at least $t(t+1)(t+2)$ full lines.

Theorem 7.10. For t = 2, every threshold symmetric design F_2 is quasi-affine and the point-set of an F_2 can be partitioned in many different ways. For every $t \geq 3$, there are at least two quasi-affine designs F_t, if one is known to exist. In particular, for every $\lambda \geq 3$, there exist at least two non-isomorphic quasi-affine designs F_λ provided λ is a prime power.

The proof of the first sentence in Theorem 7.10 can be found in [130]. The paper [136] contains a discussion of strongly quasi-affine designs, gives various necessary and sufficient conditions and proves the second part of Theorem 7.10 by showing the existence of a quasi-affine design which is not strongly quasi-affine. We also note the following from that paper:

Fact 7.11. All the known threshold symmetric designs are quasi-affine.

In a very interesting coincidence, the threshold symmetric designs F_λ play an important role in the extension problem of the infinite class of symmetric designs other than the Hadamard 2-designs appearing in the statement of Cameron's Theorem (Theorem 1.29). An attempt to make a systematic study of these extendable symmetric designs and related objects was made in Sane, S.S. Shrikhande and Singhi [141] and here we give an account of the main ideas in that paper. To that end, we first define the following four objects:

Convention 7.12. For a fixed $\lambda = 1, 2, \ldots$, let D_λ, E_λ, F_λ and Q_λ denote some t - (v, k, λ*) designs with the following specifications:

(i) D_λ denotes a symmetric design with $v = (\lambda + 1)(\lambda^2 + 5\lambda + 1) - 1$, $k = (\lambda + 1)(\lambda + 2) - 1$ and $\lambda^* = \lambda$.

(ii) E_λ is a quasi-symmetric 3-design with $v = (\lambda + 1)(\lambda^2 + 5\lambda + 5)$, $k = (\lambda + 1)(\lambda + 2)$ and $\lambda^* = \lambda$. The other parameters of E_λ are easily seen to be: $r = (\lambda + 1)(\lambda^2 + 5\lambda + 5) - 1$, $\lambda_2 = (\lambda + 1)(\lambda + 2) - 1$, $b = (\lambda + 1)(\lambda^3 + 8\lambda^2 + 19\lambda + 11) - 1$. Also $y = \lambda + 1$ and $x = 0$ (to see this use Cameron's Theorem).

(iii) F_λ is a symmetric design with $v = (\lambda + 1)^2(\lambda + 3)$, $k = (\lambda + 1)(\lambda + 2)$ and $\lambda^* = \lambda + 1$. Thus F_λ is a threshold design defined earlier (note: the shift in λ).

(iv) Q_λ denotes a quasi-symmetric design with $v = (\lambda + 1)(\lambda^2 + 5\lambda + 5) - 1$, $k = (\lambda + 1)(\lambda + 2)$ and $\lambda^* = (\lambda + 1)^2$. The other parameters of Q_λ are: $r = (\lambda + 1)^2(\lambda + 3)$ and $b = (\lambda + 1)^2(\lambda + 3)$. Also $x = 0$ and $y = \lambda + 1$ (to show the last two parameters count flags and double flags).

Note that the parameters of E_λ are those corresponding to a one point extension of D_λ and the parameters of Q_λ are those corresponding to a residual of E_λ. Also D_1 is a projective plane of order four and E_1 an $S(3, 6, 22)$. At this stage we make a slight detour and look at the role of F_λ. To that end, we make the following:

Definition 7.13. Let E be any quasi-symmetric design with block intersection numbers 0 and y. Let X be any block of E. The block residual E_X of E (with respect to X) is the incidence structure obtained by the deletion of X from the point-set of E. An incidence structure F will be called a block quasi-residual of E if it is isomorphic to E_X for some X.

Observe that E_X is always a 1-design with parameters set $(v - k, b - \frac{k(r-1)}{y} - 1, r - \frac{k\lambda}{y}, k)$. When is E_X a design? The answer is given in the following Theorem of Baartmans and M. S. Shrikhande [7] which is implicit in the proof of Cameron's Theorem 1.29.

Theorem 7.14. E_X is a design for some X if and only if E is a 3-design and hence is an extension of a symmetric design.

Proof. Let $F = E_X$ and G be the design whose point-set is the same as that of F and where blocks are all those blocks that intersect X in y points (clearly then the block size in G is k - y). Applying Lemma 3.23 (iii) to E and F obtains:

$(r - 1)(y - 1) = (\lambda - 1)(k - 1)$ and $(y - 1)(r - \frac{k\lambda}{y} - 1) = (\lambda' - 1)(k - 1)$, where λ' is the λ-parameter of F. Therefore $\lambda - \lambda' = \frac{k\lambda(y - 1)}{y\,(k - 1)}$.

Now compute the parameters of G to get $\lambda - \lambda' = \frac{k\lambda}{y}\frac{(k - y - 1)}{(v - k - 1)}$.

Hence we have $v = \frac{k^2 - 3k + 2y}{y - 1}$. From this, the value of r is obtained as $\frac{\lambda(k^2 - 3k + y + 1)}{(y + 1)(k - 1)}$. Since $r - 1 = \frac{(\lambda - 1)(k - 1)}{y - 1}$, we obtain $\lambda(k - y) = (k - 1)(k - y)$ where $k \neq y$. Therefore $\lambda = k - 1$. Contraction of E at a point p produces a dual 2-design whose parameter set is $(r, v\text{-}1, k\text{-}1, \lambda, y\text{-}1)$. Since $\lambda = k\text{-}1$, this 2-design is symmetric and therefore E_p is a 2-design. Hence E is a 3-design and the remaining part follows by Cameron's Theorem.

Exercise 7.15. Show that for any block X of $E = E_\lambda$, the block residual E_X at X is an F_λ. Conclude that E_λ is triangle-free (i.e., has no three mutually disjoint blocks).

Exercise 7.16. Show that Q_λ is triangle-free.

At this stage we make an obvious remark that Cameron's Theorem can be rephrased as follows: E is an extension of a symmetric design if and only if E is a quasi-symmetric 3-design with $(x, y) = (0, y)$ and in that case a derived design D is a projective plane of order ten, a

(495, 39, 3) - design, a Hadamard 2-design or $D = D_\lambda$ (in which case $E = E_\lambda$). Looking at the symmetric designs D_λ in the light of Fact 7.4 already tells us that we are in a hopeless situation for all $\lambda \geq 3$. Notwithstanding this cautious note, [141] introduced the notions of maximal arcs in designs F_λ, Q_λ, and D_λ and proved some structural results, which we now discuss.

Proposition 7.17. Suppose $F = F_\lambda$ is a block residual of some $E = E_\lambda$. Then all the lines of F are trivial.

Proof. If some line of F has size ≥ 3, then we have a point triple contained in $\lambda^* = \lambda + 1$ blocks of F_λ. But $\lambda_3 = \lambda$ in $E = E_\lambda$ and this contradiction proves the result.

Observe that in view of Fact 7.11, Proposition 7.17 sounds yet another hopeless note: <u>E_λ cannot be constructed from any of the 'known' F_λ's for $\lambda \geq 2$</u>.

Proposition 7.18. Let E be a family of k-subsets of a v-set such that any two members of E intersect in 0 or y points. Also let the number of members of E equal b, where v, b, k and y have the same values as those for E_λ (as given in Convention 7.12). Then $E = E_\lambda$, i.e., E is a 3-design.

Proof. Let $D = E_p$ be a derived incidence structure at a point p. Since $y \geq 2$, any two blocks of D intersect in $y - 1 > 0$ points and hence by Majumdar's inequality (see [15]), r_p = the number of blocks containing $p \leq v - 1$, which is the number of points of D. So $bk = \Sigma r_p \leq v(v - 1) = bk$ and the equality of these two numbers shows that $r_p = v - 1$ for all p, i.e., E is a 1-design with $r = v - 1$. Let a_i be the number of point-pairs of

D which occur in i blocks. Then the values of k - 1 and v - 1 can be used to compute Σa_i, $\Sigma i a_i$ and $\Sigma\, i(i-1)\, a_i$. This obtains $\Sigma(i-\lambda)^2\, a_i = 0$, i.e., D is a symmetric design and hence E is a 3-design with $\lambda = \lambda_3$ as asserted.

The notion of an arc and a maximal arc of an incidence structure is fairly old. For example, in Chapter VI, we have already come across arcs in projective planes. A systematic investigation of maximal arcs in the larger class of symmetric designs on the background of coding theory was perhaps first considered in Assmus and van Lint [3]. However, their definition of an arc is rather restrictive from our point of view, viz. the study of F_λ, D_λ, and Q_λ. We therefore make

Definition 7.19. [141] Let D be a (v, b, r, k, λ) design. An (α, n)-arc of D is a non-empty set A of n points of D with the property that $|X \cap A| \leq \alpha$ holds for every block X of D.

Lemma 7.20. If A is an (α, n) - arc of a (v, b, r, k, λ)-design D then $|A| = n \leq 1 + \dfrac{r(\alpha-1)}{\lambda}$ with equality if and only if every block intersects A in 0 or α points.

Proof. Fix $p \in A$ and count flags (z, Z) with $z \in A$ and $p \in Z$ in two ways.

Definition 7.21. A maximal (α, n) - arc is an (α, n) - arc A of D with $n = \dfrac{1 + r(\alpha-1)}{\lambda}$. A maximal (α, n)-arc will be simply called an α-arc. For a symmetric (v, k, λ)-design D, a block α-arc is defined dually: it is a set S of blocks of D with the property that any point of D occurs on 0 or α blocks of S. For an α-arc A of D, call a block B a secant (respectively a <u>passant</u>) of A if $|B \cap A| = \alpha$ (respectively equals 0). Note that these are the only two types.

Exercise 7.22. Show that, if D has an α-arc, then α divides $r - \lambda$. In particular, if a projective plane has a hyperoval, then its order is even.

Theorem 7.23. Let $D = D_\lambda$ (recall Convention 7.12) and let A be a $(\lambda + 1)$ - arc of D. Then the following assertions hold.

(i) $|A| = (\lambda + 1)(\lambda + 2)$ and any point $p \notin A$ is on $\lambda(\lambda + 2)$ secants and $\lambda + 1$ passants of A.

(ii) The set A' of all the blocks disjoint from A forms a (dual) block $(\lambda + 1)$ - arc. Conversely, given a block $(\lambda + 1)$ - arc A', a (point) $(\lambda + 1)$ - arc A can be obtained by reversing this procedure. Thus there is a one-to-one correspondence between $(\lambda + 1)$ - arcs and block $(\lambda + 1)$ - arcs.

(iii) Let A_1 and A_2 be $(\lambda + 1)$ - arcs and suppose A_i' is the block $(\lambda + 1)$ - arc induced by A_i, i = 1, 2. Then $|A_1' \cap A_2'| = |A_1 \cap A_2|$.

(iv) Let I(A) denote the incidence structure induced by D in A (i.e., non-empty intersections of blocks of D with A). Then I(A) is a $((\lambda + 1)(\lambda + 2), (\lambda + 1)^2(\lambda + 3) - 1, (\lambda + 1)(\lambda + 2) - 1, \lambda + 1, \lambda)$ - design with no repeated blocks.

(v) Suppose D is embedded in some E_λ. Then any block of E_λ which is not a block of D is a $(\lambda + 1)$ - arc of D.

Proof. For (i), use Lemma 7.20 with $r = k = (\lambda + 1)(\lambda + 2) - 1$ and $\alpha = \lambda + 1$; (ii) is then a consequence of (i) since every point not in A is on $\lambda + 1$ blocks of A' (while a point in A is on no block of A'). Use this to compute $|A'|$ = number of passants of A. For (iii), let $\beta = |A_1 \backslash A_2|$.

Then the number of blocks that are passants of A_2 and secants of A_1 (by (i) and a two-way counting) is $\frac{(\lambda+1)\beta}{\lambda+1} = \beta$. Hence by (ii), A_2 and A_1 have $(\lambda+1)(\lambda+2) - \beta$ common passants. But this is the number $|A_1 \cap A_2|$. Hence (iii). For (iv) observe that the block size in I(A) is $\lambda + 1$ and clearly the λ-parameter must be λ. Finally, (v) is a consequence of the fact that E_λ is a quasi-symmetric design with $x = 0$ and $y = \lambda + 1$.

Exercise 7.24. For two distinct $(\lambda + 1)$ - arcs A_1 and A_2 in D_λ, show that $|A_1 \cap A_2| < (\lambda + 1)^2$.

Theorem 7.25. Let $F = F_\lambda$ and let A be a $(\lambda + 1)$ - arc of F. Then the following assertions hold.

(i) $|A| = (\lambda + 1)^2$ and any point $p \notin A$ is on $(\lambda + 1)^2$ secants and $\lambda + 1$ passants of A.

(ii) Let A' be the set of those blocks that are disjoint from A. Then A' forms a (dual) block $(\lambda + 1)$ - arc. Conversely, given A', A can be obtained by reversing the procedure. Thus there is a one-to-one correspondence between $(\lambda + 1)$ - arcs and block $(\lambda + 1)$ - arcs of F.

(iii) Let A_1 and A_2 be two $(\lambda + 1)$ - arcs and suppose A_i' is the block $(\lambda + 1)$ - arc induced by A_i', i = 1, 2 (as given in (ii)). Then $|A_1' \cap A_2'| = |A_1 \cap A_2|$.

(iv) Let I(A) be the incidence structure induced by F in A. Then I(A) is a $((\lambda + 1)^2, (\lambda + 1)^2(\lambda + 2), (\lambda + 1)(\lambda + 2), \lambda + 1, \lambda + 1)$ - design.

(v) Suppose F is a block residual and hence can be embedded in some E_λ (by adding an extra block containing $(\lambda + 1)(\lambda + 2)$ new points). Then every block of E_λ which is not in F gives a $(\lambda + 1)$ - arc of F.

(vi) For $\lambda \geq 2$, if F has an embedding in some E_λ (as a block residual), then no block of I(A) (constructed in (iv)) is repeated $\lambda + 1$ times.

Proof. Most of the assertions are proved exactly as in the proof of Theorem 7.23. To prove (v), observe that every block Y of E_λ not in F intersects every block of F in 0 or $y = \lambda + 1$ and if Y' denotes the point-subset of Y consisting of points of F alone, then $|Y'| = (\lambda + 1)(\lambda + 2) - (\lambda + 1) = (\lambda + 1)^2$, i.e., Y' is a $(\lambda + 1)$ - arc of F. For (vi) notice that every point triple in E_λ occurs in λ blocks. If a block of I(A) is repeated $\lambda + 1$ times then we obtain a point-triple of F contained in $\lambda + 1$ blocks, a contradiction.

Exercise 7.26. Let D be any symmetric (v, k, λ) - design and let N be a set of points of D with the property that $|N| = k$ and $|N \cap X| \geq \lambda$ for every block X of D. Then N is a block of D.

Theorem 7.27. Consider Q_λ and let A be a $(\lambda + 1)$ - arc of Q_λ. Then the following assertions hold.

(i) $|A| = (\lambda + 1)(\lambda + 1) - 1$. Any point $p \notin A$ is on $(\lambda + 1)(\lambda + 2)$ passants of A and $(\lambda + 1)^2(\lambda + 2) - (\lambda + 1)$ secants of A.

(ii) There are precisely $(\lambda + 1)^2(\lambda + 3)$ blocks passants to A and the incidence structure D(A) on these blocks is an F_λ.

(iii) Let I(A) denote the incidence structure induced by Q_λ in A. Then I(A) is a $((\lambda + 1)(\lambda + 2) - 1, \lambda + 1, (\lambda + 1)^2)$ - design.

(iv) If A_1 and A_2 are distinct $(\lambda + 1)$ - arcs, then $|A_1 \cap A_2| = \lambda$.

(v) Suppose S is a set of $(\lambda + 1)$ - arcs of Q_λ with $|S| = (\lambda + 1)(\lambda^2 + 5\lambda + 5) - 1$. Let D(S) be the incidence structure with members of S as blocks. Then D(S) is some D_λ. Further, D_λ and Q_λ are respectively derived and residual designs of a unique E_λ.

(vi) The number of $(\lambda + 1)$ - arcs of Q_λ is at most $(\lambda + 1)(\lambda^2 + 5\lambda + 5) - 1$ with equality if and only if Q_λ is a residual of some E_λ. Hence Q_λ is embeddable in an E_λ in at most one way.

Proof. All the assertions more or less follow by methods similar to the ones used in the proof of Theorem 7.23. For example, (ii) is proved as follows: Let Y be a block of D(A). Since the r parameter of D(A) is $(\lambda + 1)(\lambda + 3)$ by (i) and since $y = \lambda + 1$, it follows that there are $(\lambda + 1)^2 (\lambda + 3) - 1$ blocks of D(A) that have non-empty intersection with Y. Hence every pair of blocks intersect in $y = \lambda + 1$ and thus the dual of D(A) is a design in which both the r parameter and k parameter equal $(\lambda + 1)(\lambda + 2)$. So D(A) is symmetric and must equal an F_λ. Consider (iv). Let m be the number of points in $A_1 \setminus A_2$ and let n be the number of secants of A_1 that are also passants of A_2. Then a two-way counting (using (i) and (ii)) produces $m(\lambda + 1)(\lambda + 2) = n(\lambda + 1)$ and $m(m - 1)(\lambda + 1) = n(\lambda + 1)\lambda$. So $m = (\lambda + 1)^2$. Therefore $|A_1 \cap A_2| = \lambda$. For (v) and (vi), let S be a set of $(\lambda + 1)$ - arcs given in the assumption of (v). Make an incidence structure E whose blocks are all the blocks of Q_λ and all the blocks of S augmented by ∞. Then E satisfies (using (iv)) all the assumptions of Proposition 7.18 and therefore $E = E_\lambda$. This also shows that the blocks in S form a $D_\lambda = D$. If we have a larger set of S' of $(\lambda + 1)$ - arcs, then we obtain a $(\lambda + 1)$ - arc N which intersects (using (iv)) every block of S in λ points. Hence N is a set of $(\lambda + 1)(\lambda + 1) - 1$ points of D with the property that $|N \cap X| = \lambda$ for all the blocks X of D. Exercise 7.26 shows that N is a block of D, i.e., $N \in S$ and the proof is complete.

Theorem 7.28. (Cameron-van Lint [49]) If $D = D_\lambda$ is a derived design of some E_λ, then the dual of D is also a derived design of some E_λ.

Proof. Let $Q = Q_\lambda$ be the corresponding residual design. By Theorem 7.23 (v), the blocks of Q are $(\lambda + 1)$ - arcs of D. For every such $(\lambda + 1)$ - arc, construct a block $(\lambda + 1)$ - arc under the correspondence given by Theorem 7.23 (ii). Let Q' be the set of all such block $(\lambda + 1)$ - arcs. To prove our theorem, it is sufficient to show that given any three distinct blocks X, Y and Z with $|X \cap Y \cap Z| = m$, the number of block $(\lambda + 1)$ - arcs in Q' containing all the three blocks is $\lambda - m$. Under the one-to-one correspondence given by Theorem 7.23 (ii) and (iii), this amounts to showing that the number of $(\lambda + 1)$ - arcs in Q that are disjoint from $X \cup Y \cup Z$ is $\lambda - m$. Consider the symmetric design $F = F_\lambda$ obtained by considering all the blocks of E disjoint from X. Since X, Y and Z are blocks of D (which is a symmetric design), $|X \cap Y| = |X \cap Z| \neq 0$ and hence Y' and Z' are $(\lambda + 1)$ - arcs of F where $Y' = Y \backslash X$ and $Z' = Z \backslash X$. Since Y and Z are blocks of D, $|Y \cap Z| = \lambda$ and hence $|Y' \cap Z'| = \lambda - m$. By Theorem 7.25 (iii), the number of blocks of F disjoint from both Y' and Z' is $\lambda - m$. So the number of blocks of E disjoint from $X \cup Y \cup Z$ is $\lambda - m$ as desired.

Theorem 7.29. Let $F = F_\lambda$ and suppose F is embeddable (as a block residual) in some $E = E_\lambda$. Then the dual of F is also embeddable (as a block residual) in some E_λ.

Proof. Let M be the unique block of E disjoint from all the blocks of F and let S be the set of all the remaining blocks of E. Suppose $S' = \{Y': Y' = Y \backslash M, Y \in S\}$. Then by Theorem 7.25 (v), every member Y' of S' is a $(\lambda + 1)$ - arc of F. For every member Y' of S' construct a block $(\lambda + 1)$ - arc B (Y') under the correspondence given by Theorem 7.25 (ii). Write $B(Y) = B(Y') \cup (Y \cap M)$ for every $Y \in S$. Then $|B(Y)| = |Y|$ and let $T = \{B(Y): Y \in S\}$. Construct a new incidence structure whose points are all the points of M and all the blocks of F. The blocks of this new incidence structure E' are all the points of F, all the members of T, and M itself. Then E' has a constant block size $(\lambda + 1)(\lambda + 2)$. Also, for $Y_1 \neq Y_2$, $|B(Y_1) \cap B(Y_2)| = |B(Y_1') \cap B(Y_2')| + |Y_1 \cap Y_2 \cap M| = |(Y_1') \cap (Y_2')| + |Y_1 \cap Y_2 \cap M|$ by Theorem 7.25 (iii). Clearly the R·H·S. equals $|Y_1 \cap Y_2| = 0$ or $y = \lambda + 1$ and hence $|B(Y_1) \cap B(Y_2)| = 0$ or y. If we look at two points of F, then they are commonly contained in $\lambda^* = \lambda + 1 = y$ blocks. Finally, according to Theorem 7.25 (ii), the number of blocks of F containing a given point of F and contained in a given B (Y) is $\lambda + 1 = y$. Therefore E' is an incidence structure satisfying all the conditions of Proposition 7.18. Therefore E' is some E_λ and certainly the dual of F is embedded in E' as described.

Observe that Theorems 7.28 and 7.29 are combinatorial versions of 'algebraic polarity' of D_λ and F_λ respectively. In fact, (a) the duality given by Theorem 7.25 (ii) and (b) if A is a 3 - arc of $F = F_2$ which is a block residual of some E_2, then all the lines of F are trivial (given by Proposition 7.17) were used in a recent paper of Bagchi [8] to show that the GF(3) code of the (transpose of the) incidence matrix of such an F must have dimension at least 16. Using this result and the fact that the trivial (complete) 3-design with all the 3-subsets of a 12-set generates a code of dimension 11 over GF(3), it can be shown that the GF(3) code generated by the incidence matrix of an E_2 must have dimension at least 27. Since E_2 is quasi-symmetric this code C is self-dual, i.e., is contained in its dual. Bagchi [8] first rules out the possibility of dimension 27 and then shows by a long argument that C cannot have dimension 28. Since the number of points of E_2 is 57, the dimension of C is at most $\left[\frac{57}{2}\right] = 28$ and that enables Bagchi [8] to prove

Theorem 7.30. E_2 does not exist. Consequently no symmetric (56, 11, 2)-design has an extension.

The authors recently learned [9] that Bagchi's paper contains an error. But fortunately it can be corrected and a corrigendum is to appear .

Exercise 7.31. Show that Theorem 7.30 is equivalent to the assertion of the non-existence of an SR graph $NL_3(18)$ on 324 vertices (with parameters given in Chapter IV).

Hint: Use Exercise 4.10 and Theorem 4.11.

We conclude this chapter by noting that the duality between point and block arcs of D_λ derives itself from the construction and uniqueness of a projective plane of order four (D_1, in our notation) by G. Higman. In this construction, which is described in detail in Cameron and van Lint [49], a projective plane of order four is constructed from its hyperoval and dual hyperoval (in the setting of vertices, edges, factors, and factorizations of the complete graph K_6). A generalization of that idea is given in the following theorem and we refer to [141] for a proof.

Theorem 7.32. The existence of a $D = D_\lambda$ with a $(\lambda + 1)$ - arc is equivalent to the existence of two designs D' and D" with parameters $v^* = (\lambda + 1)(\lambda + 2)$, $b^* = (\lambda + 1)^2(\lambda + 3) - 1$, $r^* = (\lambda + 1)(\lambda + 2) - 1$, $k^* = \lambda + 1$ and $\lambda^* = \lambda$ with no repeated blocks such that the following conditions are satisfied: Let B' and B" denote the set of blocks of D' and D" respectively. Let C" denote the set of all subsets of size $\lambda(\lambda + 2)$ of B". Then there exists a mapping $\theta : B' \rightarrow C''$ satisfying the following:

(i) For any block X' in B', $\theta(X')$ contains each point of D'' exactly λ times.

(ii) For two distinct blocks X', Y' in B' with $|X' \cap Y'| = i$, $|\theta(X') \cap \theta(Y')| = \lambda - i$.

(iii) For every block $X'' \in B''$ there exist $\lambda + 2$ blocks $X' \in B'$ such that $X'' \in \theta(X')$.

VIII. QUASI-SYMMETRIC 2-DESIGNS

In our earlier chapters, we have looked at various classes of (sometimes only parametrically possible) quasi-symmetric designs. This chapter is devoted to the study of quasi-symmetric designs in general, particularly from a structural point of view. While Cameron's Theorem (Theorem 1.29) certainly boosted the interest in quasi-symmetric designs, it seems to be only in the last ten years or so that the structural investigations of quasi-symmetric designs began. The investigations are far from complete and we wish to give an account of the work in this area. The strong regularity of the block graph of a q.s. (= quasi-symmetric) design has been exploited in a paper of Neumaier [117]. However, the design-structural properties of q.s. designs were probably first studied in a paper of Baartmans and M.S. Shrikhande [6]. In that paper, q.s. designs with $(x, y) = (0, y)$, $y \geq 2$ and with no three mutually disjoint blocks were studied. Typical examples are E_λ and Q_λ of Chapter VII (Convention 7.12). Various modifications and improvements of the results in that paper have been obtained and the results have been generalized in two different directions. The main part of this chapter is to give a summary of most of the results in these directions.

Throughout this chapter, D denotes a quasi-symmetric (q.s.) design with parameter set (v, b, r, k, λ) and with block intersection numbers x and y where $x \leq y$. The block graph G of D has parameters n, a, c and d, where $n = b$. We also let r' and s' denote the eigenvalues of Γ other than a, where $r' \geq 0 \geq s'$ is assumed in view of Theorem 2.22 and Exercise 3.10. We write f and g for the multiplicities of r' and s' respectively. If $x < y$, then D is called a proper q.s. design and D is called improper if $x = y$.

Exercise 8.1. Show that D is improper if and only if D is a symmetric design.

<u>The implicit assumption for most of the results in this chapter is: D is a proper q.s. design.</u>

Proposition 8.2. [139] Let $y < k$. Then the following assertions hold.

(i) $x < \frac{k^2}{y} < \lambda$.

(ii) If $x = 0$, then $y \leq \lambda$ and $y = \lambda$, if and only if D is improper.

(iii) Let $v \geq 2k$. Then $x \leq \frac{k}{2}$.

(iv) For a fixed value of k there exist finitely many q.s. designs (with block size k) and $x > 0$ (but x is not fixed).

Proof. By Theorem 3.8, $a = \frac{k(r-1) - x(b-1)}{y - x}$ (where the meaning of a is already explained). Since D is proper, $a \neq 0$ and hence $x < \frac{k(r-1)}{b-1} < \frac{kr}{b} < \frac{k^2}{v}$. But $r \geq k + 1$ and therefore $\lambda(v - 1) \geq k^2 - 1$. This obtains $\frac{k^2}{v} < \lambda$ proving (i). For (ii), note that $r \geq k$ by Fisher's inequality with equality if and only if D is symmetric. Use this in Proposition 3.17(i). For (iii) use (i) to obtain $x \leq \frac{k}{2}$. Since $y < k$, no block is repeated and hence by Theorem 3.15, $b \leq \binom{v}{2}$. Therefore (iv) will be proved, if we show that v is bounded. Since $x \neq 0$, (i) implies $v < \frac{k^2}{x} \leq k^2$ and our proof is complete.

Exercise 8.3. Show that the dual of a quasi-symmetric design is an SPBIBD (defined in Chapter IV).

Observe that Exercise 8.3 is essentially proved by using the following fact: Given any flag (p, B), the number of blocks containing p (respectively not containing p) that intersect B in y points is a constant d' (respectively e'). Translating this property in the set-up of strongly regular graphs, Neumaier [117] made the following:

Definition 8.4. A non-empty subset R of an SR graph Γ is called a regular set of valence d' and nexus e' if the number of vertices of R adjacent with any vertex w of Γ is d' (respectively e') if $w \in R$ (respectively $w \notin R$).

Exercise 8.5. Show that the complement of a regular set R is also a regular set R" of valence d" = a - e' and nexus e" = a - d'. Further show that the incidence structure whose points are the vertices of R and whose blocks are the vertices not in R (with incidence defined by adjacency) is a 1-design.

Exercise 8.6. Recall Definition 4.30 and Theorem 4.32. Suppose Γ is the point-graph of an SPBIBD, say E and let B be any block of E. Show that the set of points of B forms a regular set of Γ and compute its valence and nexus.

As we observed before making Definition 8.4, the block graph of a q.s. design has many (in fact, one for every set of blocks containing a given point) regular sets. Although the eigenvalues of an SR graph Γ are zeros of a monic quadratic (Theorem 2.22), the eigenvalues are not necessarily integers. However, if Γ contains a regular set R, then its adjacency matrix A can be written in the form

$$A = \begin{pmatrix} M_1 & C \\ C^t & M_2 \end{pmatrix}$$

where M_1 (respectively M_2) is the adjacency matrix of the subgraph induced on R (respectively the complement of R). Then the properties of R imply that the column vector $[(a - d')\hat{j}, -e'\hat{j}]^t$, where the first $\hat{j}$ has order $|R|$, is an eigenvector of A with the corresponding eigenvalue $d' - e' < a$. This gives

Proposition 8.7. (Neumaier [117]) The eigenvalues of an SR graph Γ with a regular set are integers. In particular, the eigenvalues of the block graph of a q.s. design D are integers.

Compare Proposition 8.7 with Corollary 3.9. Neumaier [117] uses Proposition 8.7 to define a positive and negative regular set. He also shows that

Theorem 8.8. The parameters v, b, r, k, λ, x and y of a q.s. design D are expressible in terms of the parameters r', s' (the eigenvalues) of the block graph Γ, f, g their respective multiplicities and the number of vertices n of Γ.

Neumaier's paper [117] contains some equations satisfied by the parameters of an SPBIBD and interesting inequalities. He also classifies q.s. designs in four standard classes. These are standard examples of q.s. designs given in Chapter III and will not, therefore, be repeated here. However, we must point out that much of the recent work of Calderbank [40], [41] and Tonchev [172] is motivated by Neumaier's table of 'exceptional' (those that do not belong to the four standard classes) parameter sets of q.s. designs. These papers, in addition, rule out some parameter sets occurring in [117]. In his paper [117], Neumaier calls a quasi-symmetric 2-design D exceptional, if neither D nor its complement belong to any of the above four classes. This paper contains a table of all feasible exceptional parameters on $v^* < 40$ points, $k^* < v/2$. We give below Neumaier's table.

Neumaier's Table

Quasi-symmetric 2-(v^*, k^*, λ^*) designs with intersection numbers x, y , and block graph parameters n, a, c, d ; subgraphs induced by a point with valency d', nexus e' in the block graph. The list covers all exceptional parameters with $2k^* < v^* < 40$.

No.	Ex?	v^*	k^*	λ^*	x	y	n	a	c	d	d'	e'
1	?	19	9	16	3	5	76	45	28	24	25	18
2	?	20	10	18	4	6	76	35	18	14	21	14
3	?	20	8	14	2	4	95	54	33	27	27	18
4	?	21	9	12	3	5	70	27	12	9	15	9
5	?	21	8	14	2	4	105	52	29	22	26	16
6	yes	21	6	4	0	2	56	45	36	36	15	12
7	yes	21	7	12	1	3	120	77	52	44	33	22
8	?	22	8	12	2	4	99	42	21	15	21	12
9	yes	22	6	5	0	2	77	60	47	45	20	15
10	yes	22	7	16	1	3	176	105	68	54	45	28
11	yes	23	7	21	1	3	253	140	87	65	60	35
12	?	24	8	7	2	4	69	20	7	5	10	5
13	?	28	7	16	1	3	288	105	52	30	45	20
14	yes	28	12	11	4	6	63	32	16	16	16	12
15	?	29	7	12	1	3	232	77	36	20	33	14
16	yes	31	7	7	1	3	155	42	17	9	18	7
17	?	33	15	35	6	9	176	45	18	9	27	15
18	?	33	9	6	1	3	88	60	41	40	20	15
19	?	35	7	3	1	3	85	14	3	2	6	2
20	?	35	14	13	5	8	85	14	3	2	8	4
21	yes	36	16	12	6	8	63	30	13	15	15	12
22	?	37	9	8	1	3	148	84	50	44	28	18
23	?	39	12	22	3	6	247	54	21	9	27	12

This table of Neumaier has resulted in much recent interest and activity on quasi-symmetric 2-designs, which we shall discuss in Chapter X .

We conclude the discussion of [117] with the following result of Neumaier.

Theorem 8.9. Let D be a q.s. design and let d' be the nexus of a regular set induced by a point p of D in the block graph Γ. That is, for a flag (p, B), d' is the number of blocks containing p that intersect B in y points (this number is a constant since D is quasi-symmetric). Suppose $A = (v - 1)(v - 2)$, $B = r(k - 1)(k - 2)$, and $C = r\,[d'\,(y - 1)(y - 2) + (r - 1 - d')(x - 1)(x - 2)]$. Then $B(B - A) \leq AC$ with equality if and only if D is a 3-design.

Proof. Fix a point p of D and for distinct points q and z other than p, let λ_{qz} denote the number of blocks containing p, q, and z. Summing over all q, z obtains: $\Sigma\, 1 = A$, $\Sigma\, \lambda_{qz} = B$ and $\Sigma\, \lambda_{qz}\,(\lambda_{qz} - 1) = C$. So the average of λ_{qz} is $\bar{\lambda} = B/A$ and $0 \leq (\lambda_{qz} - \bar{\lambda})^2 = (C + B) - 2\,\bar{\lambda}\,B + \bar{\lambda}^2\,A = \dfrac{AC - B\,(B - A)}{A}$. Hence the inequality follows because A is positive. Also, D is a 3-design if and only if $\lambda_{qz} = \bar{\lambda}$ for all q, z which is true if and only if $AC = B(B - A)$.

When is a (known) strongly regular graph, a block graph of some quasi-symmetric design? The answer is not known in general. However, in some cases the parametric characterization of an SR graph can be used to obtain nice results. This point is illustrated in the following theorem of Haemers [70].

Theorem 8.10. Let $s \leq m - 2$ and let $m \geq 4$. If a quasi-symmetric design with parameters $v = \frac{(m-1)(m-2)}{2}$, $k = \frac{s(m-2)}{2}$, $\lambda = \frac{s(sm-2s-2)}{2(m-3)}$ and with block intersection numbers $x = \frac{s(sm-4s+2)}{2(m-3)}$, $y = \frac{s(s-1)}{2}$ exists, then $(m-2)^{m(m-1)/2}\,[2s(m-3)(m-s-1)]^{m(m+1)/2}$ is the square of an integer.

Proof. Calculation of the other parameters r and b and the use of Theorem 3.8 tells us that the eigenvalues of the block graph Γ are $\frac{(m-2)(m-3)}{2}$, 1 and $-(m-3)$. Since the trace of the adjacency matrix is zero, the spectrum of Γ can be computed and Theorem 2.22 shows that Γ has parameters of the complement of a triangular graph T(m) of Example 2.12. For $m = 8$, the assertion is vacuously true and for $m \neq 8$, Theorem 2.26 implies that Γ is isomorphic to the complement $\overline{T}(m)$ of T(m). Let $N = [N_1 : N_2]$ be the incidence matrix of the q.s. design D in question, written in such a way that (the columns of) N_1 correspond to a copy of $\overline{T}(m-1)$ in $\overline{T}(m)$. The situation is analogous with that of Example 5.21. Then N_1 has $\binom{m-1}{2}$ rows and columns and is therefore a square matrix. If A_1 represents the adjacency matrix of the block graph corresponding to N_1, then we have $N_1^t N_1 = kI + yA_1 + x(J - I - A_1)$. The eigenvalues of A_1 are $\frac{(m-3)(m-4)}{2}$, 1 and $-(m-4)$ (in fact, just replace m by $m - 1$ in the eigenvalues of $\overline{T}(m)$) and the multiplicity of the first eigenvalue is 1. Looking at the trace of A_1, we obtain the multiplicities of the other two eigenvalues. These are $\frac{(m-1)(m-4)}{2}$ and $m - 2$ respectively. From our basic equation, we obtain the spectrum of $N_1^t N_1$:

$$\begin{pmatrix} \frac{s^2(m-2)^2}{4} & \frac{s(m-s-1)(m-2)}{2(m-3)} & \frac{s(m-s-1)}{2(m-3)} \\ 1 & \frac{(m-1)(m-4)}{2} & m-2 \end{pmatrix}$$

Since $(\det N_1)^2 = \det(N_1^t N_1)$, the product of the eigenvalues (with respective multiplicities) is a perfect square. This gives the desired result and completes the proof.

Haemers [70] remarks that replacement of s by m-s-1 leads to the complementary parameter set and hence one may restrict to $s \leq \frac{m-2}{2}$. Further, if s = 2, then we obtain quasi-residuals of biplanes (i.e., symmetric (v, k, 2)-designs) and then Theorem 8.10 follows using Bruck-Ryser-Chowla Theorem (Theorems 1.15 and 1.16) via the Hall-Connor embedding Theorem (Theorem 1.17). Also, it is unlikely that q.s. designs as in Theorem 8.10 exist for $2 < s < m - 3$.

Observe that Theorem 8.10 depended on the peculiar parameters of the block graph of the q.s. design in question. However, most of the recent results which we will now discuss, depend only on the (general) structure of a quasi-symmetric design. As we have already seen in Chapter VII, the class of q.s. designs with x = 0 is an important class from an investigator's point of view. Use of the ubiquitous equation $(r - 1)(y - 1) = (\lambda - 1)(k - 1)$ is made in proving the following equation obtained by Sane and M.S. Shrikhande [138].

Theorem 8.11. Let D be a proper (i.e., not symmetric) q.s. (v, k, λ)-design with block intersection numbers 0 and $y \geq 2$. Write k = my (where m is an integer by Corollary 3.9). Let $\bar{c}$ be the number of triangles on any edge of $\bar{\Gamma}$, i.e., $\bar{c}$ is the number of blocks disjoint from both of two given disjoint blocks. Then m, y, $\bar{c}$ and λ satisfy the following equation:

$$(m-1)\, y\, [m\, (y + 1) - m^2 - 1]\, \lambda^2 - [(m - 1)y\, \{2m\, (y^2 + 1) - 2m^2y - (y + 1)\} - (y - 1)^2\, my\bar{c}\,]\, \lambda - (my - 1)\, y^2\, (m - 1)^2 = 0.$$

Proof. By Lemma 3.23(ii), $b = 2a - d + \bar{c} + 2$, where (b, a, c, d) is the parameter set of Γ and the values of a and d are given by Theorem 3.8 (also note that $x = 0$). This gives an equation in b, r, m, y, λ and $\bar{c}$. But $b = \frac{vr}{k}$ and $v = \frac{r(k - 1)}{\lambda} + 1$. Therefore $b = \frac{r^2\,(k - 1) + \lambda r}{kr}$ and in our equation we can replace b by this expression in r, k and λ. Finally, use Lemma 3.23 (iii) to replace r by an expression in m, y and λ (this is possible since $y \geq 2$). After simplification, this gives the desired equation. For a detailed counting argument, we refer to [138].

The 'polynomial tool' of Theorem 8.11 and many other results to follow were first used by Baartmans and M. S. Shrikhande [6], where the important case $\bar{c} = 0$ of Theorem 8.11 was handled. We postpone the discussion of that aspect of the problem for the time being and first record some interesting consequences of the quadratic in Theorem 8.11.

Theorem 8.12. For a fixed pair (m, y), $y \geq 2$ there are finitely many proper q.s. designs with block intersection numbers 0 and y and block size my.

Proof. Write the equation in Theorem 8.11 in the form $A\lambda^2 + (\alpha + B)\lambda + C = 0$, where A, B and C are appropriate integral polynomial function of m and y (which are held fixed) and $\alpha = (y - 1)^2\, my\bar{c}$. If a q.s. design with relevant parameters does exist, then this quadratic has an integral solution λ. Hence the discriminant Δ equals N^2, where N is a positive integer. But $\Delta = (\alpha + B)^2 - 4AC$ and $4AC = f(m,y)$ is a constant since m and y are fixed. We have $(\alpha + B + N)(\alpha + B - N) = f(m, y)$ and therefore the number of possibilities for the positive integer $\alpha + B$ is limited by the number of factors of $f(m, y)$. Since B is a constant, α has finitely many possibilities. So λ assumes finitely many values. The proof is then completed using Lemma 3.23 (iii) and the usual parameter conditions.

Corollary 8.13. For a fixed value of the block size k, there are finitely many proper q.s. designs with block size k and with block intersection numbers 0 and $y \geq 2$.

Proof. Write $k = my$ and use Theorem 8.12.

We record that Corollary 8.13 is false if $y = 1$ is also included. Corollary 8.13 and Proposition 8.2 (iv) yield the following result.

Theorem 8.14. For a fixed value of the block size k, there are finitely many proper quasi-symmetric designs with intersection numbers $x \geq 0$ and $2 \leq y \leq k - 1$ (where x and y are not fixed).

Theorem 8.15 For a positive integer pair $(\overline{e}, z)$, let $S(\overline{e}, z)$ denote the set of all the proper q.s. designs D such that: $(x, y) = (0, y)$, where $2 \leq y \leq z$ and given any block the number of blocks disjoint from it is $\overline{n} \leq \overline{e}$. Then $S(\overline{e}, z)$ is a finite set.

Proof. Let $T(\overline{c}, y)$ denote the set of all the proper q.s. designs D with $(x, y) = (0, y)$ and with the property that given any pair of disjoint blocks, the number of blocks disjoint from both is $\overline{c}$. Then $S(\overline{e}, z)$ is contained in the union of all $T(\overline{c}, y)$, $0 \leq \overline{c} \leq e$ and $2 \leq y \leq z$. So it is enough to show that $T(\overline{c}, y)$ is finite for a fixed $(\overline{c}, y)$. By Theorem 8.12, it suffices to show that m is bounded. Suppose not, i.e., assume that there are infinitely many m's (with the same $(\bar{c}, y)$) for which a proper q.s. design (with stipulated $\bar{c}$ and y) exists. Dividing the L.H.S. of the equation in Theorem 8.11 by $m^2(m - 1)y$ gives a function $g(m, \lambda)$ with coefficients in y and $\bar{e}$ and we must necessarily have $g(m, \lambda) = 0$ by assumption. But then $\lim_{m \to \infty} g(m, \lambda) = -\lambda^2 + 2y\lambda - y^2 = -(\lambda - y)^2 = 0$ and hence $\lambda \to y$ as $m \to \infty$, a contradiction since we assumed that D is proper. This contradiction shows that there are finitely many possibilities for m and our proof is complete.

Remark 8.16. Let $y \geq 2$, m and $\bar{c}$ have the same earlier meaning for a q.s. design with block intersection numbers 0 and y. Theorems 8.12 and 8.15 show that, if (m, y) or $(y, \bar{c})$ is held fixed, then we have finitely many q.s. designs. Such a result is not true for the pair $(m, \bar{c})$ as the following exercise shows (use affine designs of Remarks 1.25).

Exercise 8.17. Let q be a prime power. Show that with $(m, \bar{c}) = (q, q - 2)$ there exist infinitely many q.s. designs (with stipulated values of m and $\bar{c}$). In fact, y can be taken to be q^i, $i = 0, 1, 2, \ldots$.

For a fixed $\lambda \geq 2$, let G_λ denote the class of <u>all</u> q.s. designs, proper or improper (i.e., symmetric designs). It would be nice to know whether G_λ is finite but that cannot be done in view of the following open conjecture of M. Hall Jr [74].

Conjecture 8.18. For a fixed $\lambda \geq 2$ there are only finitely many symmetric designs (i.e., improper q.s. designs) whose 'λ - value' is the given λ.

Fact 7.4 indicates a strong support of Conjecture 8.18. In fact the following is due to Singhi.

Conjecture 8.19 (Singhi's conjecture [138]) For a fixed $\lambda \geq 2$, G_λ is finite.

Interestingly however, Theorems 8.11 and 8.12 prove the equivalence of these two conjectures:

Theorem 8.20. For a fixed $\lambda \geq 2$, there are finitely many proper quasi-symmetric (v, k, λ) - designs with intersection numbers 0 and y (but y is not fixed). Hence Conjectures 8.18 and 8.19 are equivalent.

Proof. Since D is proper, Lemma 3.23 (iii) implies that $y < \lambda$ and $\lambda \geq 2$. Reading the polynomial in Theorem 8.11 modulo m gives: $y(\lambda - y)(\lambda - 1) \equiv 0 \pmod{m}$. Fix λ and y where $y < \lambda$. Then m divides a constant positive number and hence m has finitely many possibilities. Therefore by Theorem 8.12, D has finitely many possibilities and the proof is complete.

Essentially, Theorem 8.20 shows that Conjecture 8.18 is as hard as Conjecture 8.19 and the difficult part in proving Conjecture 8.19 is handling symmetric designs (and obtaining a bound perhaps like the one in Fact 7.4). However, as Theorem 8.20 shows, the analogue of Hall's conjecture is true for proper q.s. designs having an intersection number zero. We also observe that the entire discussion in this chapter must exclude the case $y = 1$ since in that case $\lambda = 1$ necessarily, and the existence of infinitely many Steiner systems (Remark 1.8) shows that one cannot obtain similar results in that case. Thus the case

$(x, y) = (0, 1)$ is of a very different nature. A particularly useful result in our further investigation is the following.

Lemma 8.21. [110] Let D be a q.s. design with $(x, y) = (0, y)$ and $y \geq 2$. Then $\lambda \leq k - 1$ with equality if and only if D is a 3-design.

Proof. Let D' be a contraction of D at a point p. Then the given conditions imply that the dual of D' is a q.s. design in which the λ-parameter is y-1, the k-parameter is λ and the r-parameter is k-1. Using Fisher's inequality (Theorem 1.10), $\lambda \leq k-1$ with equality if and only if (Theorem 1.12) the dual of D' is a symmetric design, i.e., if and only if D' is a symmetric design. Hence, Theorem 1.29 shows that $\lambda = k-1$ if and only if D is a 3-design.

Two major classes of quasi-symmetric designs with $(x, y) = (0, y)$ and $y \geq 2$ are known. These are the Hadamard 3-designs (with $y = \frac{k}{2}$) and the affine designs used in Remark 8.16. The third parametrically possible class is that of the 3-designs E_λ and their quasi-residuals Q_λ discussed in Chapter VII. In fact, E_1 and Q_1 are the only known objects and were constructed in Chapter VI. Mysteriously enough, other than the affine designs, the examples D in the other two classes have the following property: The complement $\overline{\Gamma}$ of the block graph Γ has no triangles, i.e., D has no three mutually disjoint blocks. If D is a Hadamard 3-design, then this is obvious since the complement of every block is also a block (Remark 1.24) and if D is E_λ or Q_λ, then this is easily seen from Exercises 7.15 and 7.16. It is therefore very important to concentrate on the subclass consisting of those q.s. designs with $(x, y) = (0, y)$, $y \geq 2$ and with no three mutually disjoint blocks. That was done by Baartmans and M.S. Shrikhande [6] and it also seems to be the first instance of the use of the 'polynomial tool' of Theorem 8.11.

Theorem 8.22. Let D be a proper triangle free (i.e., D has no three mutually disjoint blocks) quasi-symmetric design with $(x, y) = (0, y)$. Write $k = my$ (using Corollary 3.9). Then the following assertions hold.

(i) For $y \geq 2$, the following diophantine equation is satisfied:
$[m^2 - m(y + 1) + 1]\lambda^2 - [2m^2y - 2(y^2 + 1)m + (y + 1)]\lambda + y(m - 1)(my - 1) = 0.$

(ii) For a fixed value of $y \geq 2$, the function $f(m) = -4m^2[my - (y^2 + y + 1)] - 4m(y + 1) + 1$ is non-negative and is a square of an integer.

(iii) $2 \leq m \leq y + 1$

(iv) $m = 2$ (i.e., $k = 2y$) if and only if D is a Hadamard 3-design.

(v) For $y \geq 2$, $m = y + 1$ if and only if D is some E_{λ^*} or its quasi-residual Q_{λ^*} (where $\lambda^* = y - 1$).

Proof. We assume that $y \geq 2$ and leave the case $y = 1$ (which occurs only in (iii) and (iv)) to the reader as an exercise. Then Theorem 8.11 is applicable with $\bar{c} = 0$. Simplification of the resulting expression leads to the diophantine equation as in (i). The conclusion (ii), depends on the tedious computation of the discriminant of the equation in (i), treating it as a quadratic in λ and removal of all the perfect square terms. Since the design exists the discriminant must be a perfect square (because the equation in (i) has an integer solution λ). For (iii), observe that, if $m \geq y + 2$, then $f(m)$ is negative. Also $m = 1$ leads to $y = k$, i.e., D has repeated blocks and since D is proper (with $x = 0$) we must have $y = k = v$, a contradiction. Hence $m \geq 2$ and (iii) is proved. Consider (iv) and let $m = 2$, i.e., $k = 2y$. Since $y - 1$ and $k - 1$ are coprime Lemma 3.23 (iii) implies that $y - 1$ divides $\lambda - 1$ and since D is proper $y \neq \lambda$. So $\lambda \geq 2y - 1 = k - 1$ and Lemma 8.21 shows that $\lambda = k - 1 = 2y - 1$, i.e., D is a 3-design. Then $r = 4y - 1$ by Lemma 3.23 (iii). The other parameters of D can be now computed to see that D is a Hadamard 3-design.

Consider (v) and let $m = y + 1$, i.e., $k = y^2 + y$. Then (i) gives $(\lambda - (y^2 + y - 1))(\lambda - y^2) = 0$, i.e., $\lambda = y^2 + y - 1$ or $\lambda = y^2$. In the first case use Lemma 8.21 to conclude that D is a 3-design. By Lemma 3.23 (iii), $r = (y + 2)(y^2 + y - 1) + 1$. Use this to obtain v and then $\lambda^* = \lambda_3 = \frac{\lambda_2 (k - 2)}{v - 2} = y - 1$, where $\lambda_2 = \lambda$. By Convention 7.12, $D = E_{\lambda^*}$. Finally, let $\lambda = y^2$. Then Convention 7.12 tells us that D is a Q_{λ^*}, where $\lambda^* = y - 1$. This completes proof of Theorem 8.22.

Exercise 8.23 Complete the proof of Theorem 8.22 by showing that D is a proper triangle-free q.s. design with $(x, y) = (0, 1)$ if and only if D is an affine plane of order two.

No examples of triangle-free quasi-symmetric designs with $(x, y) = (0, y)$ and $3 \leq m \leq y$ seem to be known. In fact, for $m = y$ we have the following result of M.S. Shrikhande [154].

Proposition 8.24. Let D be a proper triangle-free q.s. design with $(x, y) = (0, y)$ and $m = y \geq 2$ (i.e., $k = y^2$). Then D is one of the following two types:

(i) $y = 2$, $k = 4$ and D is the 3-(8, 4, 1) Hadmard 3-design.

(ii) $y = 6$, $k = 36$, $\lambda = 15$, and $v = 232$.

Proof. Substitution of $m = y$ in the expression of Theorem 8.22 (ii) shows that $4y^2(y + 1) - 4y(y + 1) + 1$ is a square of an integer which must necessarily be odd. Hence for some integer s, $4y^2(y + 1) - 4y(y + 1) + 1 = (2s + 1)^2$ which obtains $(y + 1)y(y - 1) = s(s + 1)$. Using

Mordell's Theorem [115], the only integer solutions of this equation with $y \geq 2$ are $y = 2$ and $y = 6$. If $y = 2$, then $k = 4$ and D is proper. So Lemma 8.21 implies that $\lambda = 3$ and the same lemma also implies that D is a 3-design. It is then easy to see (using Lemma 3.23(iii)) that D is as in (i). If $y = 6$, then Theorem 8.22 (i) (with the substitution $m = y = 6$) shows that $\lambda^2 - \lambda - 210 = 0$. Since λ is positive, we obtain $\lambda = 15$. Again use of Lemma 3.23 (iii) gives r and hence $v = 232$. This shows that D is as in (ii).

In view of our discussion so far, the following definition from [105] can now be made.

Definition 8.25. Let D be a proper triangle-free q.s. design with $(x, y) = (0, y)$, and $y \geq 2$. Then D is called exceptional if $3 \leq m \leq y$ and D is called critical if $m = y + 1$.

A computer search in [6] showed that for $y \leq 199$ there are only three exceptional parametrically feasible q.s. designs. These have $(v, k, \lambda; y)$ equal to (232, 36, 15; 6), (5290, 345, 56; 23), and (1174581, 13770, 392; 162). Of these the first is given by Proposition 8.24. The first two parameter sets do not give q.s. designs. This was shown by M.S. Shrikhande [153] and is an immediate consequence of Calderbank's results [40], [41]. We postpone a discussion of these results to Chapter X. On the background of these facts, we now give an account of structural investigations of exceptional triangle-free quasi-symmetric designs carried out by Limaye, Sane, and M.S. Shrikhande [105].

For a non-flag (p, B) of D, let $\alpha(p, B)$ denote the number of blocks containing p that do not intersect B. It is clear that in a q.s. design D with $(x, y) = (0, y)$, $\alpha = \alpha(p, B) = r - m\lambda$. We call α the order of the quasi-symmetric design D. In some sense, α is a measure of the quasi-symmetricness or 'deviation from symmetricness' of D because

Exercise 8.26. Show that $\alpha \geq 0$ with equality if and only if D is a symmetric design for which λ divides k.

Proposition 8.27. The following assertions hold.

(i) $\alpha \leq \lambda + 1$.

(ii) $\alpha \leq \lambda$ if and only if D is critical.

(iii) $\alpha = \lambda$ if and only if D is some $Q_{\lambda*}$ for some $\lambda*$ (see Convention 7.12).

(iv) $\alpha = \lambda + 1$ if and only if D is some $E_{\lambda*}$ for some $\lambda*$.

Proof. Suppose $\alpha \leq \lambda$. Then $r \geq (m + 1)\lambda$. Using Lemma 3.23 (iii), $(my - 1)(\lambda - 1) \geq (y - 1)[(m + 1)\lambda - 1]$ which simplifies to $y(m - 1) \leq \lambda(m - y)$. Since the R.H.S. is positive, Theorem 8.22 (iii) implies $m = y + 1$ and hence D is critical. By Theorem 8.22 (iv) and (v), a critical D is either $Q_{\lambda*}$ or $E_{\lambda*}$. In the first case (use Convention 7.12) $\alpha = \lambda$ (and conversely) and in the second case $\alpha = \lambda + 1$ (and conversely). This proves all the assertions.

Theorem 8.28. Let D be a proper triangle-free q.s. design with $m \geq 3$. Then:

(i) $f(\alpha) = (m - 1)\alpha^2 - (m\lambda + m - 1)\alpha + (m - 1)m\lambda = 0$.

(ii) $m \leq y + 1 \leq \lambda$.

(iii) $m \leq \alpha$.

(iv) $\lambda = \dfrac{(m - 1)\,\alpha\,(\alpha - 1)}{m\,(\alpha - (m - 1))}$.

Proof. Fix a block X of D and let D_X be the block residual at X (Definition 7.13). Since D is triangle-free, the dual of D_X is a design

with parameters: $v' = b - m(r - 1) - 1$, $b' = v - k$, $r' = k$, $k' = r - m\lambda = \alpha$ and $\lambda' = y$. Let Y be a block disjoint from X and make a two-way counting of all the blocks of D to obtain: $b - 2 = 2m(r - 1) - m^2\lambda$. Substitute this value in v' to get $v' = m(\alpha - 1) + 1$. Now use the relation $b'k' = v'r'$ to find v. This gives $v = \dfrac{[m\alpha - m + 1 + \alpha]\, my}{\alpha}$.
Next, use the relation $vr = bk$ of D to obtain (put $r = m\lambda + \alpha$): $v = \dfrac{[m^2\lambda + 2m\alpha - 2m + 2]\, my}{m\lambda + \alpha}$. Now equate the two expressions for v.
After simplification, (i) is obtained. The relation (ii) is a direct consequence of Theorem 8.22 (iii), Lemma 3.23 (iii) and the fact that D is not symmetric. If $\alpha = m - 1$, then (i) gives $\alpha = 1$, i.e., $m = 2$, a contradiction since $m \geq 3$. So $\alpha \neq m - 1$ and (iv) is obtained from (i) by solving for λ. Finally, the denominator of the R.H.S. in (iv) must be positive and hence $\alpha \geq m$ which obtains (iii). This completes our proof.

Theorem 8.29. For a fixed value of any of the parameters v, b, r, k, λ, y and m such that $y \geq 2$, $m \geq 3$, there exist finitely many triangle-free q.s. designs (with $x = 0$) such that the distinguished parameter takes on a stipulated fixed value.

Proof. If y is fixed, then Theorem 8.22(iii) shows that m has finitely many possibilities and then Theorem 8.12 yields the result. Alternatively, use Theorem 8.15 with $(\bar{c}, y) = (0, y)$. By Theorem 8.28 (ii) it is now enough to show that the assertion holds for a fixed $m \geq 3$. Fix $m \geq 3$ and let Δ_1 be the discriminant of the quadratic $f(\alpha)$ in Theorem 8.28 (i). Since the quadratic has an integral solution α, $\Delta_1 = N^2$ for some non-negative integer and an easy manipulation shows that $\Delta_1 = M^2 - g(m)$, where $M = m\lambda + (m - 1) - 2(m - 1)^2$ and $g(m) = [(m - 1) - 2(m - 1)^2]^2 - (m - 1)^2$. Therefore g(m) equals $M^2 - N^2$. Since m is fixed, so is $g(m)$. Therefore the number of pairs (M, N) is limited by the number of factors of $g(m)$. So M and hence λ has finitely many possibilities. Using Theorem 8.28(ii), y has finitely many possibilities and we are done by the observations made earlier in the proof.

Exercise 8.30. Prove the following:

(i) For $m = 3, 4$, and 5 there are no exceptional triangle-free q.s. designs.

(ii) For $m = 6$, the exceptional q.s. design determined in Proposition 8.24 is the only parametrically possible example.

The remaining major part of the paper of Limaye, Sane, and M.S. Shrikhande [105] examines the quadratic $f(\alpha)$ of Theorem 8.28 (i) and makes an analysis of the zeros of $f(\alpha)$. Notice that both the zeros are positive and their product is $m\lambda$. Call the order α of D, large if $\alpha > m\lambda$ and small if $\alpha < m\lambda$.

Exercise 8.31. Show that if $m \geq 3$, then $\alpha \neq m\lambda$ and hence the above definition makes sense.

We finally summarize a few important results from [105]:

Theorem 8.32. (i) α is small if and only if $m + 3 \leq \alpha \leq 2m - 3$.

(ii) $\alpha = m + 3$ if and only if $\alpha = 2m - 3$ and these conditions are equivalent to $m = y = 6$ (of Proposition 8.24).

(iii) If m is a prime power, then the order α of D is large.

(iv) The parameters of an exceptional triangle-free q.s. design are uniquely determined by specifying m and y alone.

(v) If the block size k is a prime power, then $k = 2^n$, $n \geq 2$ and D is a Hadamard 3-design.

In view of Theorem 8.32 and our earlier discussion, we conclude here by stating the following conjecture made in [105]:

Conjecture 8.33. There exist only finitely many exceptional triangle-free q.s. designs.

We have so far considered the situation of triangle-free as well as non-triangle-free quasi-symmetric designs restricting ourselves to the important constraint $x = 0$. In the triangle-free case, the situation $x \geq 0$ was handled by M.S. Shrikhande [149], where similar 'finiteness' conclusions were obtained but under some added restrictions on the block size. The method is again that of 'polynomial tool' approach but the calculations are more tedious. In this connection, observe that almost all the results in this chapter (that depend on 'polynomial tool') heavily depend on the ability of symbolic manipulations on a polynomial expression in several variables. The computer software package MACSYMA was therefore used in Meyerowitz et. al. [114] to improve some earlier results and also give some new results. To that end, recall that $m = \frac{k - x}{y - x}$ and introduce $s = \frac{r - \lambda}{y - x}$, $z = y - x$, where m, s and z are integers by Corollary 3.9 and the fact that D is proper. Our basic equations are those given in Lemma 3.23 (i) and (ii).

Lemma 8.34. The following two quadratic equations in λ hold for any quasi-symmetric design.

(i) $[m(m-1)k]\lambda^2 + [s(m^2z(1-2x) - z) + (1-2m)(x^2-x) - m(m-1)k]\lambda + s^2xy(k-1) = 0$

(ii) $[(m-1)k]\lambda^2 + [s\{m^2(z^2+z) + m(x-2z^2) - (2x-1)z - x^2 + x\} + k(\bar{c} + 2 - m^2 - m)]\lambda - s^2zy(k-1) = 0$

Proof. Use Lemma 3.23(i), replace b by $\frac{vr}{k}$ and next v by $\frac{r(k-1)+\lambda}{\lambda}$ and simplify the resulting equation. The denominator is z^2, which can be removed to obtain equation (i). For equation (ii) make similar substitutions in Lemma 3.23 (ii) and simplify.

Theorem 8.35. In any quasi-symmetric design,

(i) $\lambda = \frac{s^2y\,(k-1)}{M}$, where $M = -m^3 + m^2s\,(z+1) + m[\,\bar{c} + 1 - s\,(z + 1 - x)] - s(x+1)$.

(ii) s satisfies the quadratic $As^2 + Bs + C = 0$, where $A = m^3y - m^2(z^2 + z) + m(1 - 2x)\,z - (x^2 - x)$, $B = -2m^4y + m^3\,(2z^2 + 2z - x) - m^2\,[(\bar{c} + 2)z^2 - (4x + \bar{c} - 1)\,z - (2\bar{c} + 3)x] + m[2x^2 - 2x - (2\bar{c} + 4)\,xz + z] - [(\bar{c} + 2)\,(x^2 - x)]$, $C = m(m^2 - \bar{c} - 1)\,[m^2y + m(x - z) - (\bar{c} + 2)\,x]$.

(iii) In (ii), $(A, B, C) \neq (0, 0, 0)$. Furthermore, the discriminant $\Delta = B^2 - 4AC$ of the quadratic in (ii) has the form $\sum_{i=0}^{7} a_i m^i$, where the a_i's are polynomial functions in x, z, $\bar{c}$, and $a_7 = -4z^2(x+z)$.

Proof. Eliminate λ^2 from (i) and (ii) of Lemma 8.34 to obtain a linear equation in λ from which (i) can be deduced. Then substitute this value of λ, in (i) of Lemma 8.34, to get the quadratic equation in (ii). This proves (ii). For all the other details, we refer to [114] where, in addition, a complete listing of the MACSYMA output is given.

The main tool in the proof of the following theorem is Theorem 8.35 and similar types of quadratic equations, a detailed analysis of the discriminant (which is very complicated and needs MACSYMA-like software package for its handling) and showing that the discriminant eventually becomes negative. We refer the reader to Meyerowitz et. al. [114] for the details.

Theorem 8.36. (i) For any fixed value of the triple $(x, y, \bar{c})$, with $0 \leq x < y$ and $y \geq 2$, and $\bar{c} \geq 0$, there are finitely many q.s. designs with intersection numbers x, y and having $\bar{c}$ triangles on any edge of $\bar{\Gamma}$.

(ii) The assertion in (i) also holds for any fixed triple (m, x, y) where $y \geq 2$.

(iii) For any fixed value of k, the assertion in (i) holds with $y \geq 2$.

(iv) All the parameters of a q.s. design with $y \geq 2$ can be expressed in terms of m, x, $z = y - x$ and $\bar{c}$.

While many examples of quasi-symmetric designs with $(x, y) \neq (0, y)$ seem to be known, the following offers an easy construction.

Exercise 8.37 Let S be the 4-dimensional projective geometry PG (4, q) over a field of order q. Let D be an incidence structure whose points are all the points of S with $v = q^4 + q^3 + q^2 + q + 1$, $k = \lambda = q^2 + q + 1$, $x = 1$ and $y = q + 1$.

Continuing our account of the results on quasi-symmetric designs with $x \neq 0$, we now look at the situation $y = \lambda$. If $x = 0$, then Lemma 3.23 (ii) shows that $r = k$, i.e., D is symmetric. Assume that D is a proper q.s. design. If $\lambda = 1$, then $(x, y) = (0, 1)$ and D is a Steiner system (and that is the only possibility). If $\lambda = 2$, then $(x, y) = (1, 2)$. Use Theorem 3.25 to prove the following.

Exercise 8.38. Show that D is a proper q.s. design with $\lambda = y = 2$ if and only if D is a residual of a biplane (i.e., a symmetric (v, k, 2) - design).

In view of Exercise 8.38, we may assume that $y = \lambda \geq 3$. Then our earlier remark shows that $1 \leq x < y$. In this situation, the case $x = 1$ was first handled by Holliday [90], where he showed that every fixed λ (with $x = 1$) determines a single infinite parametrically feasible family. He also showed that if $x = 2$ or 3, then no proper q.s. design with $y = \lambda$ can exist. The following generalization is contained in [114]:

Theorem 8.39. There is no proper q.s. design with $y = \lambda$ and x a prime.

We mention the following recent general result of Meyerowitz [113].

Theorem 8.40. For any fixed pair (x, y) with $2 \le x < y$, there are finitely many proper q.s. designs with $y = \lambda$. Hence for a fixed $\lambda \ge 3$, there are finitely many proper q.s. designs with $x \ge 2$ (but not fixed) and $y = \lambda$.

The paper of Meyerowitz [113] also contains a classification of q.s. design with $y = \lambda$ under additional restrictions. In this connection, we should also mention a recent elegant result of Tonchev [173], where a q.s. (56, 16, 6)-design with (x, y) = (4, 6) is constructed (note that $\lambda = y = 6$) and is shown to be embeddable in a symmetric (78, 22, 6)-design.

At this point, we turn back to $x = 0$ and consider the affine designs, an important class of quasi-symmetric designs for which a recent result of Cameron [46] attempts a parametric characterization. Recall from Chapters I and III (Remarks 1.26 and Example 3.6) that affine designs are quasi-symmetric with intersection numbers 0 and y for which parallelism (i.e., 'being disjoint or equal') is an equivalence relation on the set of blocks and every equivalence class of blocks partitions the point-set. This is equivalent to the assumption that given any non-flag (p, B), there is a unique block containing p that is disjoint from B. Let D be a quasi-symmetric design with (x, y) = (0, y). Earlier in this chapter (Theorem 8.22), we looked at the situation where D is triangle-free, i.e.,

D has no three mutually disjoint blocks or equivalently $\overline{\Gamma}$ is triangle-free. Cameron [46] looks at the other extreme: Let $\overline{\Gamma}$ have a clique of size as large as possible, i.e., D has many mutually disjoint blocks. Clearly this number is at most $\frac{v}{k}$ and D is said to have a spread if D has a set of blocks partitioning its point-set. Evidently then a spread has $\frac{v}{k}$ blocks and in that case $\overline{\Gamma}$ has a clique of size $\frac{v}{k}$ (which is the largest possible clique-size in $\overline{\Gamma}$). Examples are, of course, the affine designs.

Exercise 8.41. Let D be a q.s. design with $(x, y) = (0, y)$ and suppose $k = my$. Let $v = mk = m^2y$ and assume that D has a spread. Under these assumptions, show that: (i) Every block not in the spread intersects all the m blocks in the spread (in y points each). (ii) Given any point p and any block B in the spread such that p is not incident with B, there is a unique block containing p which is disjoint from B. (iii) $r = m\lambda + 1$. (iv) D is an affine design.

Theorem 8.42. (Cameron [46]) Let D be a quasi-symmetric (v, k, λ)-design with $(x, y) = (0, y)$ such that D possesses a spread. Then one of the following holds.

(i) $\lambda = 1$ (so also $y = 1$).

(ii) D is affine, i.e., $v = m^2y$, $k = my$, $\lambda = \frac{my - 1}{m - 1}$ and D possesses a parallelism such that every parallel class partitions the point-set of D.

(iii) $v = y(2y + 1)(2y + 3)$, $k = y(2y + 1)$ and $\lambda = y(2y - 1)$.

Proof. Assume that (i) or (ii) is not true. Then $\lambda \geq y \geq 2$. Suppose the given spread has t blocks. Then $v = tk$ and $k = my$. So any block not in the spread must intersect $\frac{k}{y} = m$ blocks in the spread. So $t \geq m$. If $t = m$, then Exercise 8.41 shows that D is an affine design, i.e., D is as in (ii). Therefore assume that $t = m + z$ with $z \geq 1$. The proof is broken down into ten steps.

Step 1: $\lambda \leq my - 1$.

Proof. Use Lemma 8.21

Step 2: $m - 1$ divides λz.

Proof. Let D_1 be the incidence structure whose points are the blocks in the spread and whose blocks are all the blocks not in the spread. Incidence is defined by non-empty intersection. Then D_1 is a design with parameters $v_1 = t = m + z$, $k_1 = m$ and $\lambda_1 = m^2$. So $k_1 - 1$ divides $\lambda_1(v_1 - 1)$, i.e., $m - 1$ divides $m^2\lambda(m + z - 1)$ which gives the desired result.

Step 3: $\lambda[(m - 1)^2 - mz (y - 1)] = (m - 1)(my - 1)$.

Proof. Lemma 3.23 (iii) gives $(\lambda - 1)(my - 1) = (r - 1)(y - 1)$ and we also have $r(my - 1) = r(k - 1) = \lambda(v - 1) = \lambda(myt - 1) = \lambda(my(m + z)-1)$. Eliminate r from these two equations.

Step 4: $z(y - 1) \leq m - 2$.

Proof. By step 3, $(m -1)^2 > mz(y - 1)$ and hence $m^2 - 2m \geq mz (y - 1)$.

Step 5: $\lambda = mu + 1$, u a positive integer.

Proof. Read the equation in step 3 modulo m and note that $\lambda \neq 1$ since $y \geq 2$.

Step 6: $u \leq y - 1$.

Proof. Use step 6 and step 1.

Step 7: $m - 1$ divides $z(u + 1)$.

Proof. Use step 5 and step 2.

Step 8: $u = y - 1$.

Proof. By steps 4 and 7, $z(y - 1) \leq m - 2 \leq z(u + 1) - 1$. If $u \leq y - 2$, then $z(y - 1) \leq z(y - 1) - 1$, a contradiction. So $u \geq y - 1$ and step 6 gives the desired result.

Step 9: $m = zy + 1$.

Proof. By steps 7 and 8, $m - 1$ divides zy and $m - 1 \geq z(y - 1) + 1 > zy/2$ since $y \geq 2$. So $m - 1 = zy$.

Step 10: $z = 2$.

Proof. In step 3, substitute the values of λ, m and u using steps 5, 9 and 8 respectively and simplify. This gives (after cancelling y and $y - 1$), $(z - 2)(zy - 1) = 0$. So $z = 2$. Substitution of $z = 2$ in step 9 gives $m = 2y + 1$ and then steps 8 and 5 give $\lambda = (2y + 1)(y - 1) + 1 = y(2y - 1)$. This completes the proof of the result.

Remark 8.43. The case $y = 1$ in Theorem 8.42 is realized by the Kirkman's schoolgirl problem; PG(3, 2) is an example which is both well-known and resolvable. Cameron [46] also makes an attempt to pin down the structure in the case $y = 2$. In that case, $(v, k, \lambda) = (70, 10, 6)$ and a binary code generated by the incidence matrix is considered. We conclude by noting that no example in Theorem 8.41 (iii) seems to be known except in the case $y = 1$.

We now return to the critical designs (arising at $m = y + 1$) in Theorem 8.22: these are E_λ and Q_λ (of Chapter VIII). Is it possible to parametrically characterize these quasi-symmetric designs by merely making an assumption on k or λ but deleting the triangle-free hypothesis? The recent results of Sane and M.S. Shrikhande [140] show that this can be achieved under some reasonable regularity assumption. From this point on, till the end of this chapter, assume that D is a q.s. design with $(x, y) = (0, y)$ and $y \geq 2$. We need the following property introduced by Cameron [47] in the special case when D is a symmetric design.

Property (0) D has property (0) if there is a constant $p \geq 1$ such that every triple of points is contained in 0 or p blocks.

In view of property (0), call a point-triple good if it is contained in p blocks and bad otherwise (i.e., there is no block containing it).

Exercise 8.44. Let D have property (0) and suppose D_u^* is the dual of the derived D_u at a point u. Show that if D is not a 3-design, then D_u^* is a proper q.s. design with $k^* = \lambda$, $x^* = 0$, and $y^* = p$. Hence show that if D is not a 3-design, then p divides λ. Also show that if y = 2, then D has property (0).

Theorem 8.45. Let D be proper and assume that $k = y^2 + y$. Then $\beta = \frac{\lambda - 1}{y - 1}$ is an integer and $(y + 2 - \beta)(y + 1 - \beta) = \frac{(y + 1)\bar{c}\lambda}{y}$, where $\bar{c}$ = number of triangles on an edge of $\bar{\Gamma}$ (the same meaning as in Theorem 8.11). Also, the following assertions hold.

(i) $\beta \leq y + 2$ with equality if and only if D is some E_{λ^*} (see Convention 7.12).

(ii) $\beta = y + 1$ if and only if D is some Q_{λ^*}.

(iii) If D satisfies property (0), then D is some E_{λ^*} or the unique q.s. (21, 6, 4) - design with y = 2.

Proof. Since $k = y^2 + y$, Lemma 3.23 (iii) implies that y - 1 divides $\lambda - 1$. Write $\lambda - 1 = \beta(y - 1)$, where β is an integer and $\beta \geq 2$, since D is proper. Substitute m = y + 1 and $\lambda = \beta(y - 1) + 1$ in the quadratic equation of Theorem 8.11 and simplify. This gives the desired equation. If $\beta = y + 2$, then $\lambda = (y + 2)(y - 1) + 1 = y^2 + y - 1$ and Lemma 8.21 is applicable to show that D is a 3-design and hence D is E_{λ^*} by Theorem 1.29. Alternatively, observe that $\beta = y + 2$ implies $\bar{c} = 0$ and Theorem 8.22 (v) is applicable to show that D is some E_{λ^*}. This proves (i). Similarly $\beta = y + 1$, iff $\bar{c} = 0$ and then $\lambda = (y + 1)(y - 1) + 1 = y^2$ and Theorem 8.22 (iv) implies that D is some Q_{λ^*} and (ii) is proved. Consider (iii). Then application of Lemma 3.23 (iii) to the dual of

D_u^* in Exercise 8.44 implies $(k - 2)(p - 1) = (\lambda - 1)(y - 2)$. Substitution of $k = y^2 + y$ and $\lambda = \beta(y - 1) + 1$ gives $(y + 2)(p - 1) = \beta(y - 2)$. If $y = 2$ or equivalently if $p = 1$, then (i) implies that $\beta = 3$ or 4 and we are done by (i) and (ii). So let $y \geq 3$. If y is odd, then $y - 2$ and $y + 2$ are coprime and therefore $y + 2$ divides β and again we are done by (i). If y is even, then the g.c.d. of $y + 2$ and $y - 2$ divides 4. So $y + 2$ divides $4\beta \leq 4(y + 2)$ by (i). Hence $\beta = y + 2$, $\frac{y+2}{2}$, $\frac{3(y+2)}{4}$ or $\frac{y+2}{4}$. In the first case, we are done by (i). In the second case, the basic equation (given in the statement of the present theorem) implies that $y + 1$ divides 1, a contradiction. In the third and the fourth cases, the basic equation implies $y + 1$ divides 3, i.e., $y = 2$, contrary to the assumption. Hence the proof.

Theorem 8.46. Suppose $\lambda = y^2$. Then D is some Q_{λ^*}

Proof. In view of Theorem 8.45(ii), it is enough to show that $k = y^2 + y$, i.e., $m = y + 1$. Use Lemma 3.23 to get $r = my^2 + (m - 1)y$. Next, $\lambda(v - 1) = r(k - 1)$ implies that y divides $m - 1$. Write $m - 1 = \alpha y$, α an integer, to obtain $r = y^2(\alpha y + \alpha - 1)$ and $v = (\alpha y + \alpha + 1)(\alpha y^2 + y - 1) + 1$. Using $bk = vr$ we see that $k = my$ divides vr and hence $m = \alpha y + 1$ divides $\alpha - 1$. So either $\alpha = 1$ or $\alpha y + 1 \leq \alpha - 1$ which is impossible. Therefore, $\alpha = 1$ and hence $m = y + 1$. So we are done (by Theorem 8.45 (ii)).

Theorem 8.47. Let D satisfy property (0) and suppose $\lambda = y^2 + y - 1$. Then D is some E_{λ^*} or D is a (100, 12, 5) - q.s. design with $y = 2$.

Proof. If D is a 3-design, then Lemma 8.21 is applicable and $k = \lambda + 1 = y^2 + y$. Therefore we are done by Theorem 8.45 (i). Also $p = 1$ iff $y = 2$ and we treat this case separately. So assume that D is not a 3-design, $p \neq 1$ and $y \geq 3$. Application of Lemma 3.23 (iii) to the dual D_u^* of Exercise 8.44 gives $(k - 2)(p - 1) = (y + 2)(y - 1)(y - 2)$. Since y divides k, y divides $2(p + 1)$. Also, Lemma 8.21 implies that $\lambda < k - 1$ and hence $p + 1 < y$. So $y = 2(p + 1)$. Therefore $\lambda = 4(p + 1)^2 + 2(p + 1) - 1$. By Exercise 8.44, p divides λ and hence p divides 5. Since $p \neq 1$, $p = 5$ and this gives $y = 12$, $\lambda = 155$, and $k = 387$. But then $y = 12$ divides $k = 387$, a contradiction.

We are now left with $(p, y) = (1, 2)$. Then $\lambda = y^2 + y - 1 = 5$. By Lemma 3.23 (iii), $r = 4k - 3$, so $5(v - 1) = (4k - 3)(k - 1)$, i.e., 5 divides $(k - 2)(k - 1)$. Using $vr = bk$, we obtain k divides 24. Again, Lemma 8.21 implies $k > \lambda + 1 = 6$. So $k = 8, 12$, or 24 and 5 divides $(k - 2)(k - 1)$. Hence $k = 12$ is the only possibility and D is then a q.s. (100, 12, 5)-design with $y = 2$.

Theorems 8.45, 8.46 and 8.47 as well as a number of other results in this and earlier chapters dealt with the parametric characterizations of the curious designs E_λ and Q_λ (Convention 7.12) appearing in Cameron's theorem (Theorem 1.29). It would, however, be nice to have characterizations of these designs and similar ones by certain regularity properties of the configurations. This was initiated by Mavron and M.S. Shrikhande [110] for quasi-symmetric designs with $(x, y) = (0, 2)$ and was recently generalized by Sane and M.S. Shrikhande [140] to $(x, y) = (0, y)$, for $y \geq 2$. We end this chapter by giving an account of these results. As before, our standing assumption is that D is a quasi-symmetric design with intersection numbers 0 and $y \geq 2$. The following three regularity conditions on D were given in [110].

Property (I): D has property (I), if for any given set $\{q_1, q_2, q_3, q_4\}$ of four points, the conclusion (q_2, q_3, q_4) is a good triple (see property (0)) follows from the hypothesis: (q_1, q_2, q_4) and (q_1, q_3, q_4) are both good triples.

Property II: D has property (II) for a non-flag (s, B), if given any set $\{q_1, q_2, q_3\}$ of three points contained in B, the conclusion (q_2, q_3, s) is a good triple follows from the hypothesis: (q_1, q_2, s) and (q_1, q_3, s) are both good triples.

Property (III): D has property (III) for a non-flag (s, B), if given any set $\{q_1, q_2, q_3\}$ of three points contained in B, the conclusion (q_2, q_3, s) is a bad triple follows from the hypothesis: (q_1, q_2, s) and (q_1, q_3, s) are both bad triples.

Trivially, D is a 3-design if and only if every triple is a good triple. Also, as pointed out in Exercise 8.44, $y = 2$ if and only if D has property (0) with $p = 1$. Since property (I) is a global property, we have:

Proposition 8.48. D has both the properties (0) and (I) if and only if D is a 3-design.

Proof. Let $\{q_1, q_2, q_3\}$ be a point triple. We show that it is good. Let B_{12} and B_{13} be respectively two blocks containing $\{q_1, q_2\}$ and $\{q_1, q_3\}$. If $B_{12} = B_{13}$, then we are done. If not, then $y \geq 2$ implies that these blocks intersect in a point say $s \neq q_1$. Then $\{q_1, s, q_2\}$ and $\{q_1, s, q_3\}$ are both good triples and therefore $\{q_1, q_2, q_3\}$ is also a good triple by property (I).

Lemma 8.49. [110] k divides $y(\lambda - y)(\lambda - 1)$.

Proof. From $bk = vr$ and $r(k - 1) = \lambda(v - 1)$, we have $r^2 \equiv r\lambda \pmod{k}$. From $(r - 1)(y - 1) = (k - 1)(\lambda - 1)$ of Lemma 3.23 (iii), $r(y - 1) \equiv (y - \lambda)$

(mod k). So $(y - \lambda)^2 \equiv r^2(y - 1)^2 \equiv r\lambda(y - 1)^2 \equiv \lambda(y - 1)(y - \lambda)$(mod k). Therefore $(y - \lambda)[\lambda(y - 1) - (y - \lambda)] \equiv 0$ (mod k) which yields the desired result.

Theorem 8.50. [110] Let y = 2 and suppose D has property (III). Then one of the following holds.

(i) D is the extension of a projective plane of order 2 or 4.

(ii) D is the symmetric (7, 4, 2)-design.

(iii) D is the unique q.s. (21, 6, 4)-design, i.e., $D = Q_1$ (in Convention 7.12).

Proof. Let D have property (III) for the non-flag (s, B). Define a relation ~ on the points of B by: $i \sim j$ if $i = j$ or if $i \neq j$ and (s, i, j) is a bad triple. By property (III), ~ is an equivalence relation on B. For $i \in B$, the number of points not equivalent to i is λ (use y = 2) and hence the equivalence class of i has size $k - \lambda$. So $k - \lambda$ divides k and therefore also divides λ. If $\lambda = 2$, then the proof is easy and we obtain (ii). Let $\lambda \geq 3$, then Lemma 8.49 implies that k divides $2(\lambda - 1)(\lambda - 2)$. But $k - \lambda$ divides both k and λ and therefore $k - \lambda$ divides 4. Thus $k - \lambda = 1, 2$, or 4. In the first case Lemma 8.21 and Theorem 1.29 imply that D is as in (i) (because y = 2). $k = \lambda + 4$ is ruled out by Lemma 8.49 and other basic equations. Let $k = \lambda + 2$. Then Lemma 8.49 implies that $\lambda + 2$ divides $2(\lambda - 1)(\lambda - 2)$, i.e., $\lambda = 4, 6, 10$, or 22. Out of these $\lambda = 6, 10$, or 22 are ruled out by the basic equations. If $\lambda = 4$, then k = 6 and D is Q_1 (as in (ii)).

We now borrow the definition of an arc from Chapter VII. An arc in a q.s. design D with y = 2 is a set A of points with the property that A contains no good triple. Recall also that a biplane is a symmetric (v, k, 2) - design.

Exercise 8.51. Show that if $y = 2$ and if A is an arc then $|A| \leq \frac{r}{\lambda} + 1$ with equality if and only if every block intersects A in 0 or 2 points.

An arc A with $|A| = \frac{r}{\lambda} + 1$ will be called a <u>maximal arc</u>.

Theorem 8.52. Let D be a q.s. design with $(x, y) = (0, 2)$ such that D has a maximal arc. Then one of the following holds.

(i) D is a biplane with k even.

(ii) D is a proper q.s. design with $v = (s + 1)^2(s^2 + 3s + 1) + 1$, $k = (s + 1)(s + 2)$ and $\lambda = s + 3$, $s \geq 1$.

Proof. By Exercise 8.51, λ divides r and therefore Lemma 3.23 (iii) implies that λ divides $k - 2$. Write $k = s\lambda + 2$, $s \geq 1$. If $s = 1$, then $k = \lambda + 2$ and it is easily seen that property (III) holds for every non-flag (with each equivalence class of size 2). Since D <u>has</u> bad triples, D is <u>not</u> a 3-design. Hence by Theorem 8.50, D is the unique symmetric (7, 4, 2) - design or the unique (21, 6, 4) - q.s. design (which are in (i) and (ii) respectively). Let $s \geq 2$. We have $\lambda \geq y = 2$ with equality implying that D is a biplane as in (i). Let $\lambda \geq 3$ and $s \geq 2$. Use Lemma 3.23 (iii) to obtain $r = (s\lambda - 1)(\lambda - 1) + 1$, then find v and use $vr = bk$ to obtain $(s + 2)(\lambda - 2) \equiv 0 \pmod{s\lambda + 2}$, where $\lambda \geq 3$. Hence $\lambda \geq s + 3$. But $(s + 2)(\lambda - 2) \equiv 2(\lambda - (s + 3)) \pmod{s\lambda + 2}$. Therefore $s\lambda + 2$ divides $2(\lambda - (s + 3))$. If $\lambda \neq s + 3$, then $(s - 2)(\lambda + 2) \leq -12$, which is absurd. So $\lambda = s + 3$ and this gives D as in (ii).

Note that in Theorem 8.52 (ii), $s = 2$ gives a (100, 12, 5) q.s. design with $(x, y) = (0, 2)$ whose existence is not known. We end this chapter by recording the following theorem based on property (II). This theorem generalizes a result of Mavron and M.S. Shrikhande and its proof can be found in [140].

Theorem 8.53. Let D be a q.s. design with $(x, y) = (0, y)$, $y \geq 2$. Assume that D has property (0) and property (II) for some non-flag. Then D is one of the following four types.

(i) $y = 2$ and D is the unique 3-(8, 4, 1) Hadamard 3-design.

(ii) D is a biplane and if D has a characteristic, then the characteristic is three.

(iii) D is some E_{λ^*}.

(iv) D is a (100, 12, 5) - q.s. design with $y = 2$.

Remarks 8.54. We mention the papers of M.S. Shrikhande [151] for an exposition of the (quadratic) polynomial tool method in quasi-symmetric designs and [153] for a recent survey on quasi-symmetric designs, intersection numbers and codes. Although there has been much activity in the area of quasi-symmetric designs, the classification problem for general quasi-symmetric 2-designs appears to be very difficult. However, the situation for quasi-symmetric 3-designs is perhaps more optimistic. This is the topic of the next chapter.

IX. TOWARDS A CLASSIFICATION OF QUASI-SYMMETRIC 3-DESIGNS

All through the development of this monograph up to the present point (especially Chapters V through VIII), we have looked at various properties and characterizations, both parametric and geometric, of quasi-symmetric designs, essentially to the point of convincing ourselves that the problem of determination of all the quasi-symmetric designs is indeed a hard problem. With that background, this chapter will show us that the situation for quasi-symmetric 3-designs is more promising. This is to be expected since a derived design of a q.s. 3-design at any point is also quasi-symmetric and that gives us more information (Example 5.24). However, quite unlike the case of q.s. 2-designs, very few q.s. 3-designs seem to be known. In fact, up to complementation there are only three known examples of q.s. 3-designs with $x \neq 0$. Application of the 'polynomial method' to this rather mysterious situation (see Cameron [44]) is the theme of the present chapter.

Cameron's Theorem (Theorem 1.29) has been one of the focal points of our study of q.s. designs. This is more so for q.s. 3-designs since the extensions of symmetric designs obtained in that theorem are in fact quasi-symmetric with $x = 0$ (and conversely). Despite its completeness, the classification given by Cameron's theorem is perhaps and unfortunately only parametric as revealed in the following discussion. A Hadamard 3-design exists if and only if a Hadamard matrix of the corresponding order exists. Hadamard matrices are conjectured to exist for every conceivable (i.e., a multiple of four) order. The existence of Hadamard 3-designs is therefore linked directly with the Hadamard matrix conjecture. The existence of a (112, 12, 1)-design has been settled in the negative by Lam et al. [101].

Nothing is known about a 3-(496, 40, 1) design, which is an extension of a symmetric (495, 39, 3)-design (cf. Fact 7.4). Finally, among the E_λ, only E_1 is known (and besides the Hadamard 3-designs, is among the three known examples of q.s. 3-designs up to complementation as stated in Fact 9.1 below) and is the 3-(22, 6, 1) design (with intersection numbers 0 and 2), E_2 does not exist by Bagchi's Theorem 7.30. Nothing is known about E_λ for $\lambda \geq 3$ and Fact 7.4 already tells us that no E_λ with $\lambda \geq 3$ can be constructed with a 'known symmetric design as a derived design.'

On the background of quasi-symmetric 3-designs with x = 0, the 'existence knowledge' of arbitrary quasi-symmetric 3-designs is no better as observed by Cameron [44] and as revealed in the following:

Fact 9.1. If E is a 'so far known' quasi-symmetric 3-(v, k, λ) design with

v > k + 2, then E is one of the following five types:

(i) E is a Hadamard 3-design (with x = 0 and $y = \frac{k}{2}$).

(ii) E is E_1 (with (x, y) = (0, 2)).

(iii) E is the 4-(23, 7, 1) design with (x, y) = (1, 3).

(iv) E is the 3-(22, 7, 4) design with (x, y) = (1, 3) (which is a residual of the design in (iii)).

(v) E is a complement of any one of the designs of type (i) through (iv).

Whether Fact 9.1 is a theorem or otherwise is difficult to predict at the moment. A weaker conjecture to that effect was made in an investigation of quasi-symmetric 3-designs by Sane and M.S. Shrikhande [139] and the main results of this chapter attempt to shed more light on that conjecture:

Conjecture 9.2. [139] Let E be any quasi-symmetric 3-design with $v > k + 2$ and with the smaller intersection number x. Then E is one of the following three types:

(i) $x = 0$ and E is one of the designs occurring in Cameron's Theorem (Theorem 1.29).

(ii) $x = 1$ and E is the Witt design (the 4-(23, 7, 1) design) or its residual the 3-(22, 7, 4) design.

(iii) E is a complement of one of the designs occurring in (i) or (ii).

The results of Lam et al. [101] and Bagchi (see Chapter VII, Theorem 7.30) do not seem to have much bearing on the validity of Conjecture 9.2. However, a recent result of Calderbank and Morton [42] and Pawale and Sane [119] shows that q.s. 3-designs with $x = 1$ are precisely those occurring in (ii) of Conjecture 9.2. Besides some recent results of Pawale [118, 120] which we discuss at the end of this chapter, the above analogue of Cameron's Theorem indicates a support of Conjecture 9.2. An important remark that must be recorded in this connection is the fact that q.s. 4-designs are completely determined by a result of Ito et al. [98] and Bremner [31]. We begin this chapter with a brief description of the tight t-designs and give a proof of the Ray-Chaudhuri and Wilson bound [129].

Theorem 9.3. [129] Let $t = 2s$, $s \geq 1$, and let D be a t-(v, k, λ) design without repeated blocks. Suppose $v \geq k + t$ and $k \geq t$. Then the following assertions hold:

(i) The number of blocks $b \geq \binom{v}{s}$.

(ii) D has at least s intersection numbers.

(iii) $b = \binom{v}{s}$ if and only if D has exactly s intersection numbers.

No elementary proof of (ii) and (iii) of Theorem 9.3 seems to be known. We content ourselves by giving a proof of Theorem 9.3(i) from Cameron and van Lint [49] which depends on the following:

Exercise 9.4. Let D be a t-(v, k, λ) design and $m + n \leq t$, where m and n are non-negative integers. Suppose $\lambda_{m,n}$ denotes the number of blocks containing a given m-tuple and disjoint from a given n-tuple. Show that $\lambda_{m,n}$ is a constant that depends only on v, k, λ (and t). Further show that if $k + t \leq v$, then $\lambda_{m,n} \neq 0$.

Proof of Theorem 9.3.(i): Construct a generalized incidence matrix N with rows indexed by the $\binom{v}{s}$ s-tuples of points and columns indexed by the b blocks (this is as in the proof of Theorem 3.15). The entries in N are 1 or 0 with entry 1 at (S, B) if and only if the s-tuple S is contained in the block B. Since the column rank of N is at most b, it suffices to show that the column rank of N is $\binom{v}{s}$. Let $\mathbf{R}^{\binom{v}{s}}$ be the $\binom{v}{s}$-dimensional vector space consisting of all the (column) vectors of size $\binom{v}{s}$ over **R**. Let γ_S be the column vector with entry 1 at S and 0 otherwise, i.e., γ_S is the characteristic vector of S. Then $\{\gamma_S$: S is an s-tuple$\}$ is a basis of our vector space and hence the assertion is proved if we can show that γ_S belongs to the column space of N for every s-tuple S.

Let ρ_B denote the column of N indexed by B. We show that each γ_S is a linear combination of ρ_B's. Let S, T etc. denote s-tuples. Fix an s-tuple S and for i = 0, 1, 2, ..., s let y_i denote $\sum \rho_B$, where the sum is taken over all the blocks B with $|B \cap S| = i$.

$$\text{Then } y_i = \sum_{j=0}^{i} \sum_{|T \cap S| = j} \sum_{\substack{B \supseteq T \\ |B \cap S| = i}} \gamma_T.$$

Fix an s-tuple T intersecting S in j points. We need to find the number of blocks intersecting S in i points and containing T. This can be done by choosing an (i - j) - tuple from S\T. After choosing and fixing such an (i - j) - tuple, B must contain exactly s + (i - j) points of

$S \cup T$ and hence must not contain the remaining s-i points of $S \cup T$. Exercise 9.4 shows that this number $\lambda_{s+i-j,\, s-i}$ is a constant.

Therefore, $y_i = \sum_{j=0}^{i} [\binom{s-j}{i-j} \lambda_{s+i-j,\, s-i}]\, x_j$, where $x_j = \sum_{|T \cap S| = j} \gamma_T$. It is now clear that y_i is a linear combination of x_j's. In fact $y_i = \sum a_{ij} x_j$, where $a_{ij} = 0$ if $j > i$ and $a_{ii} = \lambda_{s,\, s-i} > 0$, by Exercise 9.4.

Write $A = [a_{ij}]$, $i, j = 0, 1, 2, \ldots, s$, where a_{ij}'s depend on v, k and λ. Clearly A is a triangular matrix with non-zero diagonal. Hence A is non-singular and therefore the x_j's are linear combinations of the y_i's. In particular, $x_s = \gamma_s$ is a linear combination of the y_i's. Since y_i is a linear combination of ρ_B's, we have expressed γ_s as a linear combination of the columns of N and the assertion is proved.

Any t-design with $t = 2s$ and $b = \binom{v}{s}$ is called a tight t-design. Tight 2-designs are just the symmetric designs (and infinitely many of them exist). Peterson [122] has shown that there are no tight 6-designs. No tight t-design with $t \geq 8$ seems to be known and a result of Bannai [10] shows that for every fixed $t \geq 10$, there are finitely many tight t-designs. By Theorem 9.3 (iii), a tight 4-design is just a quasi-symmetric 4-design. These 4-designs are completely determined by the following result of Ito et al. [98] and Bremner [31].

Theorem 9.5. D is a tight 4-design if and only if D is the Witt 4-(23, 7, 1) design or its complement.

Using Theorem 9.3, Cameron [44] proved that a (2s + 1)-design must have at least s + 1 intersection numbers.

Exercise 9.6. List all the examples of (2s + 1)-designs with s + 1 intersection numbers.

As a matter of fact, the proof of the remaining parts of Theorem 9.3 involve a polynomial (called the Delsarte (annihilator) polynomial), whose coefficients are polynomial functions of v, k and λ and the zeros of that polynomial are the s intersection numbers. This annihilator polynomial also exists for a (2s + 1)-design with s + 1 intersection numbers and is also implicit in Delsarte's work [60]. An elementary alternative proof of the existence of that polynomial was recently given by M.S. Shrikhande and Singhi [155]. The proof given below is both shorter and more elementary than the one given in [155].

Theorem 9.7. Let D be a $(2s + 1)$ - (v, k, λ) design with s + 1 intersection numbers $x_1, x_2, \ldots, x_{s+1}$. Then there exists a polynomial f(z) of degree s + 1 with coefficients (explicit) polynomial functions of v, k and λ such that the s + 1 zeros of f(z) are $x_1, x_2, \ldots, x_{s+1}$.

Proof. Let λ_j be the number of blocks containing a given j-tuple. Then $\lambda_j = \frac{(v - j)(v - j - 1) \ldots (v - t + 1)}{(k - j)(k - j - 1) \ldots (k - t + 1)} \lambda$, where $j = 0, 1, 2, \ldots, t$. Suppose B is a fixed block and let m_i denote the number of blocks intersecting B in x_i points, $i = 1, 2, \ldots, s + 1$. Then counting all the pairs (T, C), where T is a j-tuple contained in B and $C \neq B$ produces

$$\sum_{i=1}^{s+1} m_i[x_i]_j = (\lambda_j - 1)[k]_j,\ j = 0, 1, \ldots, s + 1 \qquad (1)$$

where $[x]_r = x(x - 1) \ldots (x - r + 1)$ is the 'falling factorial.' The standard transformation (see e.g. Brualdi [35]) $x^{\ell} = \sum_{j=0}^{\ell} s(\ell, j)[x]_j$, where $s(\ell, j)$ is the Stirling number of second kind converts (1) to

$$\sum_{i=1}^{s+1} m_i x_i^{\ell} = b_{\ell},\ 0 \le \ell \le 2s+1, \tag{2}$$

where $b_{\ell} = s(\ell, j)\ (\lambda_j - 1)\ [\,k]_j$ called the 'power sum' is a number that depends on v, k and λ. Now put $f(z) = \prod_{i=1}^{s+1} (z - x_i) = \sum_{j=0}^{s+1} f_j z^{s+1-j}$, so that f_j is the j-th elementary signed symmetric function in $x_1, x_2, \ldots, x_{s+1}$ and $f_o = 1$ (i.e., $f_1 = -(x_1 + x_2 + \ldots x_{s+1})$ etc.) we have $0 = f(x_i) = \sum_{j=0}^{s+1} f_j x_i^{s+1-j}$. Therefore, for $s + 1 \le \ell \le 2s + 1$, we obtain $\sum_{j=0}^{s+1} f_j b_{\ell - j}$ $= \sum_{j=0}^{s+1} \sum_{i=1}^{s+1} m_i x_i^{\ell - j} f_j = \sum_{i=1}^{s+1} m_i x_i^{\ell-(s+1)} \sum_{j=0}^{s+1} x_i^{s+1-j} f_j =$ $\sum_{i=1}^{s+1} m_i x_i^{\ell-(s+1)} f(x_i) = 0$. Hence, the (unknown) elementary symmetric functions f_j and the (known) power sums b_j satisfy

$$\sum_{j=0}^{s+1} f_j b_{\ell - j} = 0, \text{ where } s + 1 \le \ell \le 2s + 1 \tag{3}$$

Since $f_o = 1$, it is easy to see that (3) can be rewritten in the form

$$B\,\underline{f} = \underline{b} \tag{4}$$

where $\underline{f} = [f_1, f_2, \ldots, f_{s+1}]^t$, $\underline{b} = -[b_{s+1}, b_{s+2}, \ldots, b_{2s+1}]^t$ and

$$B = \begin{bmatrix} b_s & b_{s-1} & . & . & . & . & b_0 \\ b_{s+1} & b_s & . & . & . & . & b_1 \\ \vdots & \vdots & & & & & \vdots \\ b_{2s} & b_{2s-1} & . & . & . & . & b_s \end{bmatrix}$$

Here B is a (known) matrix of power sums, the R.H.S. is known and hence f_j's are uniquely determined, if we can show that the matrix B of order s + 1 is non-singular. Clearly, the f_j's determine $x_1, x_2, ..., x_{s+1}$ uniquely (up to permutation) and hence the proof is complete, if we show that B is non-singular.

In [155], a decoding algorithm of B-C-H codes is invoked to show that B is non-singular. However, the following elementary argument obtains the same assertion. Define the column vectors

$$w_g = [m_1 x_1^g, m_2 x_2^g, \ldots, m_{s+1} x_{s+1}^g]^t$$

and $\quad z_g = [b_g, b_{g+1}, \ldots, b_{g+s}]^t$, where $g = 0, 1, 2, \ldots, s+1$.

Let A be the Vandermonde matrix

$$A = \begin{bmatrix} 1 & 1 & \cdots & 1 \\ x_1 & x_2 & \cdots & x_{s+1} \\ \vdots & \vdots & & \vdots \\ x_1^s & x_2^s & \cdots & x_{s+1}^s \end{bmatrix}$$

Then A is non-singular since the x_i's are distinct. Also $Aw_g = z_g$ follows from our basic equation (2). Hence, $Aw_o = z_o$ and $w_o = [m_1, m_2, \ldots, m_{s+1}]^t$ is uniquely determined by $x_1, x_2, \ldots, x_{s+1}$ and v, k, λ. So the m_i's are independent of the choice of B. Therefore if some $m_i = 0$, then D has at most s intersection numbers and D is also a 2s-design by Theorem 1.5 (b). But this contradicts Theorem 9.3 (ii). So $m_i \neq 0$ for all i. Now write $W = [w_0, w_1, \ldots, w_s]$ and $Z = [z_0, z_1, \ldots, z_s]$. Then both Z and W are matrices of order s + 1 and AW = Z. A simple observation shows that there is only a slight difference between A and W. In fact, $\det W = m_1 m_2 \ldots m_{s+1} \det A$ and our earlier observation shows that $\det W \neq 0$, i.e., W is non-singular. So Z is non-singular. We leave it to the reader to check that, if the columns of Z are read from right to left, then it is B, i.e., application of the permutation π with $\pi(i) = s - i$ changes Z into B. So B is non-singular as desired and we are done.

Remark 9.8. The computational aspects of the annihilator polynomial in Theorem 9.7 are far more complex than showing its existence. One relies on the recursive algorithm in Peterson and Weldon [123] or Berlekamp [14]. A result of Cameron, which we already mentioned, tells us that q.s. 5-designs do not exist and q.s. 4-designs are determined by Theorem 9.5. Therefore the explicit determination of the annihilator polynomial for q.s. 3-designs becomes particularly important. This is implicit in Neumaier's proof [117] of Theorem 8.9 and two recent proofs are due to Calderbank [39] and Sane and M.S. Shrikhande [139]. We give the proof given in the latter because it is simpler and relies on the obvious: a derived design of a q.s. 3-design is also a quasi-symmetric design whose parameters and intersection numbers are determined by those of the original extension.

Theorem 9.9. Let D be a quasi-symmetric 3-(v, k, λ) design with intersection numbers x and y, where $x \geq 1$. Then the intersection numbers are zeros of a quadratic $f(\alpha) = A\alpha^2 + B\alpha + C$, where A, B and C are (polynomial) functions of v, k and λ given by:

$A = (v - 2)[\lambda(v - 1)(v - 2) - k(k - 1)(k - 2)]$,

$-B = [(v - 2) + 2(k - 1)^2](v - 1)(v - 2)\lambda - [k(v - 2) + (k - 1)^2]k(k - 1)(k - 2)$ and

$C = (k - 1)^2 k^2 [\lambda(v - 2) - (k - 1)(k - 2)]$.

Proof. Write $\alpha_1 = x + y$ and $\alpha_2 = xy$. Then Lemma 3.23 (i) reads:

$$k(r - 1)\alpha_1 - (b - 1)\alpha_2 = k(k - 1)(\lambda_2 - 1) + k(r - 1) \qquad (1)$$

Let D' be a derived design. Since $x \neq 0$, Theorem 1.29 implies that D' is a proper q.s. design with $k' = k - 1$, $r' = \lambda_2$, $b' = \lambda_1 = r$, $\lambda' = \lambda_3$, $x' = x - 1$ and $y' = y - 1$. Apply Lemma 3.23 (i) to D' and suitably rearrange the terms to obtain

$$[(k-1)(\lambda_2 - 1) + (r-1)]\,\alpha_1 - (r-1)\,\alpha_2$$
$$= (k-1)(k-2)(\lambda_3 - 1) + 3(k-1)(\lambda_2 - 1) + (r-1) \qquad (2)$$

We then solve (1) and (2) for α_1 and α_2 using Cramer's rule. The determinant of the 2 x 2 coefficient matrix is:

$$F = \begin{vmatrix} k(r-1) & -(b-1) \\ (k-1)(\lambda_2 - 1) + (r-1) & -(r-1) \end{vmatrix}$$

$$\text{Let } G = \begin{vmatrix} k(k-1)(\lambda_2 - 1) + k(r-1) & -(b-1) \\ (k-1)(k-2)(\lambda_3 - 1) + 3(k-1)(\lambda_2 - 1) + (r-1) & -(r-1) \end{vmatrix}$$

and

$$H = \begin{vmatrix} k(r-1) & k(k-1)(\lambda_2 - 1) + k(r-1) \\ (k-1)(\lambda_2 - 1) + (r-1) & (k-1)(k-2)(\lambda_3 - 1) + 3(k-1)(\lambda_2 - 1) + (r-1) \end{vmatrix}$$

Use of Theorem 1.5 (b) shows that $F = \theta A$, $G = -\theta B$ and $H = \theta C$, where A, B, C are as in the statement of the theorem and θ equals $\dfrac{\lambda(v-k)^2(v-2)}{k(k-1)^2(k-2)^2}$. Since $\theta \neq 0$, $F = 0$ only if $A = 0$, which implies that

$\lambda_3(v - 1)(v - 2) = k(k - 1)(k - 2)$, which by Theorem 1.5 (b) implies $r = k$, i.e., D is symmetric, a contradiction. So $F \neq 0$ and the proof is complete using Cramer's rule.

Corollary 9.10. Let D be a q.s. 4-design. Then x and y are zeros of the (Delsarte) quadratic

$$f(\alpha) = \alpha^2 - \left[\frac{2(k-1)(k-2)}{v-3} + 1\right]\alpha + \lambda\left[2 + \frac{4}{k-3}\right].$$

Proof. By Theorem 1.29, $x = 0$ implies that D is an extension of a symmetric design D' which is also a 3-design, a contradiction. So $x \geq 1$ and Theorem 9.9 is applicable, where the 'λ' in Theorem 9.9 is actually λ_3. By Theorem 9.3 (iii), $b = \binom{v}{2}$, which by Theorem 1.5(b) implies $\lambda_3 = \frac{k(k-1)(k-2)}{v-2}$. Substitute this value in Theorem 9.9 to obtain the desired result.

Corollary 9.11. Let D be a quasi-symmetric 3-(v, k, λ) design and suppose $\alpha = x$ or y, $1 \leq x < y$. Then the following equation holds:

$$\lambda[(k-1)^2 (v-2)\{k^2 - 2\alpha(v-1)\} + \alpha(\alpha - 1)(v-1)(v-2)^2]$$
$$= k(k-1)(k-2)(k-\alpha)[(k-1)^2 - \alpha(v-2)].$$

Proof. Use Theorem 9.9 and suitably add a linear term in α on both sides.

Exercise 9.12. Show that in any q.s. 3-design $v - 2$ divides $k(k-1)^3 (k-2)$.

The following general result (which is essentially equivalent to Theorem 9.9) is originally due to Neumaier [117] (but in the form of

Theorem 8.9); Calderbank [39] gave a proof of that result using Hahn polynomials.

Theorem 9.13. In a q.s. 2-design D,
$xy(v-2)^2 + [xy - k(k-1)(x+y-1)](v-2) + k(k-1)^2(k-2) \geq 0$ with equality if and only if D is a 3-design.

Proposition 9.14. [139] Let D be a q.s. 3-design with $x = 1$. Then the following equations are satisfied.

(i) $$\frac{\lambda(v-2)}{k(k-2)} = \frac{(k-1)^2 - (v-2)}{k^2 - 2(v-1)}.$$

(ii) $(k-2)(\lambda - 1) = (\lambda_2 - 1)(y-2)$.

(iii) $$\lambda = \lambda_3 = \frac{(k-2)(k-y)}{(k-2)^2 - (y-2)(v-2)}.$$

(iv) $yv^2 - y(k^2 - k + 3)v + k(k-1)^2(k-2) + 2y(k^2 - k + 1) = 0$.

Proof. These are all more or less routine derivations that follows from Corollary 9.11, Theorem 1.5 (b) and Lemma 3.23 (iii). Put $\alpha = x = 1$ in Corollary 9.11 and note that $(k-1)^2 = (v-2)$ if and only if $k^2 = 2(v-1)$. In either case, $2k-1 = v$ which forces $k = 2$, a contradiction, hence (i). (ii) is a consequence of Lemma 3.23 applied to a derived design of D. For (iii) obtain the values of λ_2 from (ii) and Theorem 1.5(b) and equate them. Finally, (iv) follows by substituting λ from (iii) into (i).

We now go back to Fact 9.1 and observe that the only known non-trivial q.s. 3-design with $x = 1$ is the 4-(23, 7, 1) design or its residual. It was conjectured in [139] that this is actually the case. The conjecture was proved by Calderbank and Morton [42] using sophisticated and long arguments involving deeper notions in number theory. In fact, the hard proof of Calderbank and Morton [42] actually finds all the integer points on two elliptic curves. While the methods used in that proof are likely to prove powerful in handling the general existence problem of quasi-symmetric designs (we will see more of this in Chapter X) and Conjecture 9.2 in particular, a very short proof of the same result was given almost at the same time by Pawale and Sane [119]. We include the latter proof since besides being short it is also quite elementary.

Theorem 9.15. Let D be a quasi-symmetric 3-(v, k, λ) design with $v \geq k + 3$ and with the smaller block intersection number one. Then D is either the 4-(23, 7, 1) Witt design or its residual the 3-(22, 7, 4) design.

Proof. [119] Clearly, $y = 2$ if and only if $\lambda = 1$. Let $\lambda = 1$ and $y = 2$. Then Proposition 9.14(i) gives $2(v - 2)^2 - 2(k^2 - k - 1)(v - 2) + k(k - 1)^2 (k - 2) = 0$. Looking at the discriminant of this quadratic we find that the only non-negative solution is $v = 5$ and $k = 3$ which contradicts $v \geq k + 3$ (actually we get the trivial 3-(5, 3, 1) design). So $y \geq 3$ and the proof is now broken into five steps. If D' is a derived design, then D' is proper quasi-symmetric with $k' = k - 1$ and $r' = \lambda_2$. By Fisher's inequality, we get $k' + 1 \leq r'$ (because D' is proper). Hence $k \leq \lambda_2$. If $\lambda \leq y - 1$, then Proposition 9.14(ii) implies $k - 1 \geq \lambda_2$, a contradiction. Therefore we have:

Step I $\lambda \geq y$.

Step II $g(\lambda) = A\lambda^2 + B\lambda + C = 0$, where $A = k^2(y - 4) + 8k - 2(y + 2)$, $B = -y[k^2 (y - 4) - k (y - 10) - (y + 6)]$ and $C = -y[k^2 - k(y + 2) + 2y]$.

Proof. In Proposition 9.14(ii), substitue $\lambda_2 = \dfrac{\lambda(v - 2)}{k - 2}$ to obtain

$$v - 2 = \frac{\lambda(k - 2)^2 - (k - 2)(k - y)}{\lambda(y - 2)}.$$

Substitute this value of $v - 2$ in Proposition 9.14(i) and simplify. This obtains the desired quadratic $g(\lambda)$.

Step III If $y = 3$, then D is one of the two designs stipulated in the conclusion of the theorem.

Proof. Use the quadratic in Step II. Since $y = 3$, $y - 1 = 2$ divides $k - 1$, i.e., k is odd and the quadratic in Step II takes the form $A\lambda^2 + B\lambda + C = 0$, where $A = k^2 - 8k + 10$, $B = -3(k^2 - 7k + 9)$ and $C = 3(k - 2)(k - 3)$. The discriminant is negative for $k \geq 9$ and no integer solution is possible for $k = 3$ or 5. So $k = 7$ and we get $\lambda^2 - 9\lambda + 20 = 0$, i.e., $\lambda = 5$ or 4. Use Proposition 9.14 (iii) to get $v = 23$ or 22 respectively. In the first case notice that any 4-tuple is contained in at most one block and since the number of blocks is $b = \frac{5 \times 23 \times 22 \times 21}{7 \times 6 \times 5} = 253$ and since $\binom{23}{4} = 253 \times \binom{7}{4}$, it is easy to see that D is a 4-design, i.e., D is a 4-(23, 7, 1) design which is unique (essentially by Proposition 6.22(i)). In the other case D is a 3-(22, 7, 4) design and we invite the reader to prove (using the knowledge in Chapter VI) that such a design is a residual of the Witt design.

Step IV $y \neq 4$.

If $y = 4$, then the quadratic in Step II can be reduced to a monic quadratic $h(k) = k^2 - 2(\lambda^2 - 3\lambda + 3)k + (3\lambda^2 - 10\lambda + 8) = 0$ whose discriminant is $4\Delta_1$, where Δ_1, in view of Step I, lies strictly between the two perfect squares $(\lambda^2 - 3\lambda + 1)^2$ and $(\lambda^2 - 3\lambda + 2)^2$. So this equation has no integral solution for k.

Step V $y \geq 5$ is impossible.

Proof. Let $y \geq 5$. Look at the quadratic in Step II again and notice that A is positive while B and C are both negative. Hence we must take the larger (positive) root. If $-B > yA$, then $-(y - 2)(k - 1)$ is positive, which is not true. So $-B \leq yA$ and similarly $-C \leq yA$. Hence the positive solution of the quadratic in Step II is given by:

$$\lambda = \frac{-B}{2A} + \frac{\sqrt{\Delta_2}}{2}, \text{ where } \Delta_2 \text{ equals } \left(\frac{-B}{A}\right)^2 + 4\left(\frac{-C}{A}\right) \le y^2 + 4y.$$

$$\text{So, } \lambda \le \frac{y}{2} + \frac{\sqrt{y^2 + 4y}}{2} < y + 1.$$

By Step I, we have $\lambda \ge y$. Hence $\lambda = y$. Substitution of $\lambda = y$ in the quadratic of Step II obtains the following monic quadratic equation:

$k^2 - (y^2 - y + 2)\,k + y^2 = 0$. Since $y \ge 5$, the discriminant of this quadratic lies strictly between the two perfect squares $(y^2 - y - 1)^2$ and $(y^2 - y)^2$. Hence this quadratic equation has no integer solution for k. This completes the proof of Step V and Theorem 9.15.

Exercise 9.16. Show that a q.s. 3-(22, 7, 4) design essentially has a unique embedding (up to isomorphism) in the 4-(23, 7, 1) Witt design. Hint: Use all the knowledge of Chapter VI.

While the situation of the existence question of q.s. 3-design with $x \ge 2$ is not clear, some results have been recently proved that indicate a support to the validity of Conjecture 9.2. Many of these results (mainly due to Calderbank [39, 40] and Tonchev [173]) depend on applications of coding theory and other areas of mathematics and involve more sophisticated methods. We postpone the discussion to Chapter X. However, we now include here some recent results based on counting arguments. Among them is the analogue of Theorem 8.22 for q.s. 3-designs (with any (x, y)) proved by Pawale [118].

Theorem 9.17. Let D be a triangle-free quasi-symmetric 3-design (i.e., D has no three blocks mutually intersecting in x points). Then D is either a Hadamard 3-design (which is self-complementary) or $D = E_\lambda$ for some λ or D is a complement of some E_λ.

Exercise 9.18. Show that Conjecture 9.2 is equivalent to the assertion that every q.s. 3-design which is neither one of the designs the 4-(23, 7, 1) design nor its residual the 3-(22, 7, 4) design nor a complement of either one of them, is necessarily triangle-free or is a 3-(496, 40, 3) design.

The proof of Theorem 9.17 is based on the following two results.

Lemma 9.19. Let E be a 3-design whose residual at a point p is a symmetric design D. Then E is a complement of a q.s. 3-design E* with $x^* = 0$.

Proof. Since the blocks not containing p in E precisely give the blocks containing p in E* and since the complement of a symmetric design is symmetric, it is evident that E* is an extension of a symmetric design. Hence by Exercise 1.30, E* has $x^* = 0$.

Lemma 9.20. D is a q.s. 3-design with $v = 2k$ if and only if D is a Hadamard 3-design.

Proof. This is well-known (see e.g. [15]). Alternatively, substitute $v = 2k$ in Theorem 9.13 to get an equation from which it easily follows that $2xy = \beta k(k-1)$, where β is an integer. Substitution for xy in the same equation leads to $\beta(2k-1) = 2(x+y) - k \le 3k - 2$ and hence $\beta = 0$ or 1. If $\beta = 0$, then we are done. If not, then $x + y = \frac{3k-1}{2}$ and $xy = \frac{k(k-1)}{2}$, which can be solved to obtain $y = k$, a contradiction.

In the proof of Theorem 9.17, Pawale [118] makes use of Lemma 9.19, 9.20, and other counting arguments mainly based on Lemma 3.23(ii) (here $b = 2a - d + 2$) applied for the design and its contraction or residual, one of which must be proper quasi-symmetric. In [120], Pawale has obtained support to the validity of Conjecture 9.2. We record some of these results in Theorem 9.21 and refer the reader to [120] for a proof.

Theorem 9.21. Let D be a quasi-symmetric 3-(v, k, λ) design. Then the following assertions hold.

(i) $v - 2$ divides $k(k - 1)^2 (k - 2)$.

(ii) $$\frac{(k-1)^2 (v-k+1)}{(v-2)(v-k)} \leq x + y - 1 \leq \frac{2(k-1)(k-2)}{v-3}.$$

(iii) $$\frac{k(k-1)^2}{(v-2)(v-k)} \leq xy \leq \frac{k(k-1)^2 (k-2)}{(v-2)(v-3)}.$$

(iv) $$\frac{k}{v-k+1} \leq \frac{xy}{x+y-1} \leq \frac{k(k-1)}{2(v-2)}.$$

(v) Upper bounds in (ii), (iii) and (iv) are attained if and only if D is the Witt 4-design or its complement.

The inequality in Theorem 9.21 (ii) was first obtained by Calderbank [39] using the linear programming bound. While all the parts of Theorem 9.21 of Pawale should prove useful in deciding the validity of Conjecture 9.2, the most important result is perhaps the following.

Theorem 9.22. For a q.s. 3-design D, let $\overline{D}$ denote the complement of D and suppose that $\overline{x}, \overline{y}$ are the intersection numbers and $\overline{k} = v - k$ the block size in $\overline{D}$. Then $x + y \geq k - 1$ implies that $\overline{x} + \overline{y} \leq \overline{k}$.

Concluding Remarks 9.23. The problem of complementation, in our opinion, is a difficult problem and a possible proof of Conjecture 9.2 will probably have to base itself on a (reasonable) assumption such as $k \leq \frac{v}{2}$. It is in this context that Theorem 9.22 becomes an important result. In this connection, also refer to the proof of Theorem 9.5 given by Ito et. al. [98] and Bremner [31]. We end the discussion by adding that the final support to Conjecture 9.2 comes from a numerical evidence: A result of A.E. Brouwer and E.J.L.J. van Heyst (which is stated in Calderbank [39] ,p. 63]) shows that a computer has checked the validity of Conjecture 9.2 for $k \leq 1{,}000$.

X. CODES AND QUASI-SYMMETRIC DESIGNS

The importance of coding theory as a valuable tool in the study of designs has been known for quite some time. We mention, for example, M. Hall, Jr. [74], MacWilliams and Sloane [107], Pless [124], and also the monographs by Cameron and van Lint [49] and Tonchev [175]. Recently Tonchev [172], [174], Calderbank [40], [41], and Bagchi [8] have proved some very nice results about designs using coding theory. We have referred to Bagchi's result (Theorem 7.30) in an earlier chapter.

The paper of Tonchev [172] has shown the link between quasi-symmetric designs and self-dual codes. Calderbank [40], [41] has proved some elegant non-existence criteria about 2-designs in terms of their intersection numbers. The proof of one of Calderbank's results [40] depends on some deep theorems of Gleason and Mallows, and MacWilliams-Sloane [107] on weight enumerators of certain self-dual codes. The results of Calderbank [40], [41] and Tonchev [172] when specialized to quasi-symmetric designs give strong results about existence, non-existence or uniqueness. For example, Tonchev [174] shows the falsity of a part of the well known Hamada conjecture concerning the rank of the incidence matrix of certain 2-designs. Some results of Tonchev [172] and Calderbank [40], [41], seem to have been motivated by Neumaier's table of exceptional quasi-symmetric designs given in Chapter VIII.

The purpose of this chapter is to review some of the results of Tonchev [172], [174] and Calderbank [40], [41] which rely on codes as one of their principal tools. As applications we rule out some quasi-symmetric 2-designs encountered in Chapter VIII.

We first recall some standard terminology and results of coding theory. For more information, we refer to MacWilliams and Sloane [107] or Cameron and van Lint [49] for a concise treatment. Though codes can be defined over arbitrary fields, we will restrict ourselves to GF(2).

A binary code C of length n and dimension k, is a k-dimensional subspace of the n-dimensional vector space V_n over GF(2). We also refer to C as an (n, k) - code. The code words are written as row-vectors. For the purpose of this chapter alone, an incidence matrix of a design D is a b x v matrix N with the rows indexed by the blocks and the columns indexed by the points. Clearly N is the transpose of a usual incidence matrix defined in Chapter I.

If C is an (n, k) - code, then its dual code $C^{\perp}$ is defined by $C^{\perp} = \{y \in V_n: xy = 0$, for all $x \in C\}$. Here xy denotes the usual inner product in V_n. The code $C^{\perp}$ is an (n, n-k)-code. The code C is called self-orthogonal if C is contained in $C^{\perp}$ and self-dual if $C = C^{\perp}$. Any (0, 1) - matrix with the property that its row space generates the code C is called a generator matrix of C. Generator matrices of the dual code $C^{\perp}$ are called the parity-check matrices of C. Elements of C are called codewords and the weight of a codeword is the number of non-zero coordinates and the minimum of all the non-zero weights of a code is called the minimum-weight of the code. If an (n, k) - code has minimum weight d, then it is referred to as an (n, k, d) - code.

Exercise 10.1. Show that if a non-zero code C is defined over a field F, then C is self-orthogonal implies that F has a finite characteristic.

All through this chapter, C is the code (over GF(p) for some prime p), generated by the rows of N. Evidently, then rank (N) = dim (C) (over GF(p)) and $C^{\perp}$ has N as its parity check matrix, i.e., a (row) vector $\underline{z} \in C^{\perp}$ if and only if $N\underline{z}^t = 0$. Bearing this in mind, we also note the following well-known facts.

Result 10.2. (i) A code C has minimum weight $\geq$ d, if and only if every d - 1 columns in a parity-check matrix are linearly independent.

(ii) The weights of all codewords in a self-orthogonal binary code C are even.

If in (ii) of the above result, in addition, all code words have weights divisible by four, the code is called doubly-even. Concerning doubly-even codes, we state the following result from [124].

Theorem 10.3. A doubly-even (n, n/2) code exists if and only if n is divisible by 8.

The next lemma from Tonchev [172] shows how to produce self-orthogonal codes from certain quasi-symmetric designs.

Lemma 10.4. Let N be a b x v incidence matrix of a quasi-symmetric (v, k, λ)-design having intersection numbers x, y satisfying $k \equiv x \equiv y \pmod 2$. (i) If k is even, then the binary code of length v with generator matrix N is self-orthogonal. (ii) Let B be the bordered b x (v + 1) matrix obtained by adding an extra-column of all ones to (the left of) N. If k is odd, then B generates a binary self-orthogonal code of length v + 1.

The proof of the above lemma is immediate upon observing that the weights of all rows of the generator matrix, as well as the inner products of any two rows is always an even number. Also notice that if $k \equiv 0 \pmod 4$ (resp. $k \equiv 3 \pmod 4$), then in (i) (resp. (ii)), the code is doubly-even, i.e., all weights are divisible by four.

The next two lemmas are also from Tonchev [172].

Lemma 10.5. If N is a b x v incidence matrix of a 2-(v, k, λ) design, then the dual of the binary code with generator matrix N has minimum weight $d \geq (r + \lambda)/\lambda$, where r denotes (as usual), the number of blocks on a point.

Proof. Let S be a minimum set of linearly dependent columns of N, where N is treated as a matrix over GF(2) and suppose $|S| = m$. Then every set of m - 1 columns is linearly dependent and Result 10.2 (i) shows that $\dim (C^{\perp}) \geq m$. Suppose n_j is the number of rows that have j ones in (the columns of) S. Since the columns of S add to a zero column vector, $n_j = 0$ for odd j and we have the obvious incidence equations:

$$\sum 2in_{2i} = rm \text{ and } \sum 2i(2i - 1)n_{2i} = m(m - 1)\lambda.$$

So $\sum 2i(2i - 2)\, n_{2i} = m[(m - 1)\, \lambda - r]$ and every summand on the L.H.S. is non-negative. Therefore $m - 1 \geq \frac{r}{\lambda}$ and the assertion follows.

Lemma 10.6. If N is a b x v incidence matrix of a 2-(v, k, λ) design, then the dual of the binary code C with generator matrix B has minimum weight $d \geq \{(b + r)/r\, ,\, (r + \lambda)/\lambda\}$, where B is the bordered matrix as in Lemma 10.4 (ii).

Proof. This is similar to the proof of Lemma 10.5. Again let S be a minimum set of linearly dependent columns of B. If the left-most column vector j of all ones does not belong to S, then $|S| = m \geq (r + \lambda)/\lambda$ and we are done by Result 10.2(i). Suppose j belongs to S. Then the columns of S' add to j, where S' is the set of all the columns of S except j. Hence, if n_i is the number of rows of N that have i ones in common with the columns of S', then $n_i = 0$, for even i and $\sum n_{2i+1} = b$, and $\sum (2i + 1)\, n_{2i+1} = r\,(m - 1)$, i. e., $\sum 2in_{2i+1} = r\,(m - 1) - b \geq 0$. So $m \geq (b + r)/r$ and we are done.

Using the results of the previous lemmas, Tonchev [172] obtains the following.

Corollary 10.7. If E is a self-dual code containing the code of Lemma 10.4, then the minimum weight of E is $\geq (r + \lambda)/\lambda$ in case (i), and $\geq \min\{(b + r)/r, (r + \lambda)/\lambda\}$ in case (ii).

Proof. If C is the code of Lemma 10.4 and C is contained in E, then $E = E^{\perp}$ is contained in $C^{\perp}$, which implies that the minimum weight of $C^{\perp} \leq$ minimum weight of $E^{\perp}$.

The following result is crucial for the proofs of some results from Tonchev [172] and also Calderbank [40]. Their proofs may be found in MacWilliams and Sloane [107, Chapter 19, Section 6].

Result 10.8. Every self-orthogonal code of even length is contained in a self-dual code, and every doubly-even code of length divisible by 8 is contained in a doubly-even self-dual code of the same length.

Before proceeding further, let us take another look at Neumaier's Table of exceptional quasi-symmetric designs (See Chapter VIII). Designs No. 6, 7, 9, 10, and 11 are constructed from the Witt 5-(24, 8, 1) design. The last three were constructed by Goethals and Seidel [67] by using the derived designs of 5-(24, 8, 1) design.

Theorem 10.9. The following quasi-symmetric 2-designs with intersection numbers x, y are unique up to isomorphism:

(i) 2 - (21, 6, 4); $x = 0, y = 2$.

(ii) 2 - (21, 7, 12); $x = 1, y = 3$.

(iii) 2 - (22, 7, 16); $x = 1, y = 3$.

We reproduce below, Tonchev's proof [172] for (i) and refer to that paper for the remaining cases.

Let N be the 56 x 21 incidence matrix of a 2 - (21, 6, 4) design with $x = 0$, $y = 2$. Consider the binary code C of length 22 generated by the matrix $\begin{bmatrix} 1\ 1\ 1 \ldots & 1 \\ & 0 \\ A & 0 \\ & \vdots \\ & 0 \end{bmatrix}$. Clearly, C is self-orthogonal. The dual code $C^{\perp}$ is obtained from the code with parity-check matrix A by adding a new position equal to 0 for the words of even weight, and 1 for the words of odd weight. Hence the minimum weight d of $C^{\perp}$ is even, and since $r = 16$, by Lemma 10.5, $d \geq 6$. Then using Corollary 10.7 and Result 10.8, C is contained in a self-dual (22, 11, 6) - code. By using a result of Pless and Sloane [125], the only self-dual (22, 11, 6) code (up to a permutation of the coordinates) is the shortened Golay code G_{22}. The code G_{22} contains exactly 77 words of weight 6 forming an incidence matrix of the unique Steiner system $S_1(3, 6, 22)$. Each point in $S_1(3, 6, 22)$ is contained in 21 = 77 - 56 blocks. Hence a quasi-symmetric 2-(21, 6, 4) design must be isomorphic with a residual of the unique 3-(22, 6, 1) design.

Using the same method for the parameters of 2-(31, 7, 7) with $x = 1$, $y = 3$ and using the classification of doubly-even self-dual (32, 16, 8) codes in Conway and Pless [58], Tonchev [174] obtains the following result:

Theorem 10.10. There are exactly five isomorphism classes of quasi-symmetric 2-(31, 7, 7) designs. They all have rank 16 over GF(2).

Tonchev observed that one of the above five designs is formed by the points and planes in PG(4, 2) (c. f. Exercise 8.37). He thus provides in [174] a counterexample to the 'only if' part of the well known

<u>Hamada-Conjecture</u>: "If N(D) is the incidence matrix of a design D having the same parameters as those of the design G defined by the flats of any given dimension in PG(t, q) or AG(t, q), then $\text{rank}_q N(D) \geq \text{rank}_q N(G)$, with equality if and only if D is isomorphic to G."

In the paper [172], he also rules out designs no. 15 and 13 in Neumaier's table (see Chapter VIII). We state this result of Tonchev as

Theorem 10.11. The following two quasi-symmetric 2-designs with intersection numbers x, y do not exist.

(i) 2-(29, 7, 12); $x = 1, y = 3$.

(ii) 2-(28, 7, 16); $x = 1, y = 3$.

For a proof of the above result, we refer to Tonchev [172].

Remark 10.12. While the coding theoretic arguments used in Tonchev's proof [172] of Theorem 10.9 are interesting, we believe that they are not any easier than the (purely) combinatorial arguments for all the three parts of Theorem 10.9. In fact, Theorem 10.9 (i) is already proved in Chapter VII (Theorem 7.2). We indicate a proof in the remaining two cases leaving some verifications to the reader:

Consider a q.s. (21, 7, 12) design D with $(x, y) = (1, 3)$. This design parametrically arises as a residual of the $S_1(4, 7, 23)$ design constructed in Chapter VI (by the deletion of two points) and we will show that this is the only manner in which D could arise. A simple counting argument shows that D has $b = 120$ blocks and $r = 40$. Fix two points p and q and let a_i be the number of points s other than p and q such that the triple (p, q, s) is contained in i blocks. Then $\sum a_i = 21 - 2 = 19$, $\sum i\, a_i = 60$ and $\sum i(i - 1)\, a_i = 130$, i.e., $\sum (i - 3)(i - 4)\, a_i = 0$, which shows that every triple occurs in 3 or 4 blocks. Let (p, q, s) be a point triple contained in the four blocks B_i, $i = 1, 2, 3, 4$. Then we have exactly two points, say α

and β that are not on any B_i. Any other block C containing both p and q must intersect each B_i in a point (which is not s) and hence must contain precisely one of α and β. Since every triple is contained in 3 or 4 blocks and since we have $\lambda - 4 = 8$ such blocks C, it follows that (p, q, α) and (p, q, β) are both contained in 4 blocks each and these 8 blocks are all different. Now define for any two points p, q,

$\mathcal{L}_{pq} = \{\gamma : \gamma = p \text{ or } \gamma = q \text{ or the triple } (p, q, \gamma) \text{ is contained in 4 blocks}\}$.

Then $|\mathcal{L}_{pq}| = 5$ and our argument essentially proves $\mathcal{L}_{pq}$ is determined by any two of its members. Hence with $\mathcal{L}_{pq}$'s as lines we obtain a projective plane π of order four. Clearly, the set of blocks of D forms a class of Baer subplanes of π. It is now clear that by the addition of the two other classes of hyperovals (along with two extra points) and the lines of π to D, produces the S(4, 7, 23). The converse is trivial.

The case of a q.s. (22, 7, 16) D with (x, y) = (1, 3) is even easier: A single point, say ∞ is contained in 56 blocks and hence is not contained in b - 56 = 120 blocks which form a q.s. (21, 7, 12) design D* of the previous case. Therefore a projective plane π of order four can be defined such that π has these blocks as Baer subplanes. If a triple contained in any line of π is also contained in any of the 56 blocks containing ∞, then it is contained in 5 blocks of D, i.e., D has at least 3 + 5.4 = 23 points, a contradiction. Therefore the blocks containing ∞ form a class of hyperovals of π and the remainder of the proof is easy.

Exercise 10.13. Complete the above proof by supplying all the missing links from Chapter VI.

Recently Calderbank [40], [41] has proved some powerful non-existence criteria for 2-designs. The paper [40] deals with 2-designs whose (block) intersection numbers $s_1, s_2, \ldots, s_n$ satisfy $s_1 \equiv s_2 \equiv \ldots \equiv s_n$

(mod 2). The proofs of the main theorems of [40] depend on a result which concerns weight enumerators of certain self-dual codes. For our purpose, we need the following weaker versions of Gleason [66] and Mallows and Sloane [See 40]:

Result 10.14. (i) Let C be a self-dual binary code with all the weights divisible by 4. Then the length of C is divisible by 8. (ii) If C is an $(n, \frac{n-1}{2})$ self-orthogonal code in which all the weights are divisible by 4, then the length n of C is of the form $8m \pm 1$.

The other paper of Calderbank [41] addresses the case where $s_1 \equiv s_2 \equiv \ldots \equiv s_n \pmod p$, where p is an odd prime. These results of Calderbank provide another illustration of why coding theory is an important tool in designs. Calderbank's results when applied to q.s. designs give simple non-existence tests. Calderbank uses these tests to rule out some further cases in Neumaier's table given in Chapter VIII. We will also use these results to rule out some other q.s. designs encountered in Chapter VIII. In our opinion, these results of [40], [41] are an important new addition in design theory. For this reason, we now discuss some of the main results of Calderbank [40], [41] in more detail.

Suppose D is any 2-(v, k, λ) design with block intersection numbers $s_1, s_2, \ldots, s_n$ and with a $b \times v$ incidence matrix N. Let Γ_i $(i = 1, 2, \ldots, n-1)$ be the block graphs, where blocks E, F of D are adjacent iff $|E \cap F| = s_i$. Let A_i be the adjacency matrix of Γ_i. Then the following two observations are made in Calderbank [40]:

$$N^tN = (r - \lambda)I + \lambda J \tag{1}$$

and

$$NN^t = kI + s_1A_1 + s_2A_2 + \ldots + s_{n-1}A_{n-1} + s_n(J - \sum_{i=1}^{n-1} A_i - I) \tag{2}$$

The following lemma of Calderbank [40] weakens one of the

hypothesis of Tonchev [172] (see Lemma 10.3, this chapter). Since it is one of the crucial tools needed of [40], we also give the proof.

Lemma 10.15. let p be a prime and D a 2-(v, k, λ) design with intersection numbers $s_1, s_2, \ldots, s_n$ satisfying $s_1 \equiv s_2 \equiv \ldots \equiv s_n \equiv s \pmod{p}$ and assume that D is not symmetric.

Then,

(i) $k \equiv s \pmod{p}$,

and

(ii) $r \equiv \lambda \pmod{p}$.

Proof. All matrix equations are treated over GF(p).

(i) Suppose $k \not\equiv s \pmod{p}$. Then consideration of the matrix equation (2) for $NN^t \pmod{p}$ gives $NN^t = (k - s) I + sJ \pmod{p}$. This implies that over GF(p), NN^t has at least one non-zero eigenvalue $k - s$ repeated $b - 1$ times. Hence $v \geq \text{rank}\,(NN^t) \geq b - 1$, since D is not symmetric. Fisher's inequality forces $b = v + 1$, which using $bk = vr$ gives $k = v$, a contradiction.

(ii) Suppose $r \not\equiv \lambda \pmod{p}$. Then the matrix equation (1) tells us that $\text{rank}\,(N^tN) \geq v-1$ (exactly as in the proof of (i)). By (i), $k \equiv s$ (mod p). First assume $k \equiv 0 \pmod{p}$. Then N is a generator matrix of a self-orthogonal code C whose dimension is at most $\left[\frac{v}{2}\right]$. Hence $v - 1 \leq \text{rank}\,(N^tN) \leq \text{rank}\,(N) \leq \left[\frac{v}{2}\right]$, i.e., $v \leq 2$ which is impossible. Now let $k \not\equiv 0 \pmod{p}$. Then by suitably adding at most two columns to the left of N, one consisting of all ones and the other consisting of all the α's such that either $1 + \alpha^2 \equiv -k$ or $\alpha^2 \equiv -k \pmod{p}$ one obtains a matrix M such that M generates a self-orthogonal code C of length at most $v + 2$. Hence $\dim(C) \leq \left[\frac{v+2}{2}\right]$. But then $v - 1 \leq \text{rank}\,(N) \leq \text{rank}\,(M) \leq \left[\frac{v+2}{2}\right]$, i.e., $v \leq 4$. We leave it to the reader to handle this trivial case.

The next result of Calderbank concerns arbitrary 2-(v, k, λ) designs. It is also a crucial tool needed in [40].

Lemma 10.16. Let D be a 2-(v, k, λ) design with intersection numbers $s_1, s_2, \ldots, s_n$. Let C be the code spanned by the blocks of D and suppose $z \in C^{\perp}$. Then the following assertions hold:

(i) If weight $(z) \equiv 2 \pmod 4$, then $r \equiv \lambda \pmod 4$.

(ii) Let $s_1 \equiv s_2 \equiv \ldots \equiv s_n \equiv s \pmod 2$. If weight $(z) \equiv 1 \pmod 4$, then $r \equiv 0 \pmod 8$.

(iii) Let $s_1 \equiv s_2 \equiv \ldots \equiv s_n \equiv s \pmod 2$. If weight $(z) \equiv -1 \pmod 4$, then $2\lambda + r \equiv 0 \pmod 8$.

Proof. Let $z \in C^{\perp}$ be a code word of weight $4w + \ell$, $w \geq 1$ and $\ell = -1, 1$, or 2. Let $N_{2i} = N_{2i}(z)$ be the number of blocks meeting z in 2i points. Then we have

$$\begin{bmatrix} 1 & 1 & 1 \ldots & 1 \ldots \\ 0 & 2 & 4 \ldots & 2i \ldots \\ 0 & 1 & 6 \ldots & \binom{2i}{2} \end{bmatrix} \begin{bmatrix} N_0 \\ N_2 \\ \vdots \\ N_{2i} \\ \vdots \end{bmatrix} = \begin{bmatrix} b \\ (4w + \ell)\, r \\ \frac{1}{2}(4w + \ell)(4w + \ell - 1)\lambda \end{bmatrix}$$

Multiply this matrix equation on the left by the matrix $\frac{1}{8}\begin{bmatrix} 8 & -5 & 2 \\ 0 & 6 & -4 \\ 0 & -1 & 2 \end{bmatrix}$

to obtain $\frac{1}{8}\begin{bmatrix} 8\ 0\ 0 \ldots & 8 + 4i(i-3) & \ldots \\ 0\ 8\ 0 \ldots & -8i(i-2) & \ldots \\ 0\ 0\ 8 \ldots & 4i(i-1) & \ldots \end{bmatrix} \begin{bmatrix} N_0 \\ N_2 \\ \vdots \\ N_{2i} \\ \vdots \end{bmatrix}$

$$= \frac{1}{8}\begin{bmatrix} (4w+\ell-1)(4w+\ell)\lambda - 5r(4w+\ell) + 8b \\ 3r(4w+\ell) - (4w+\ell-1)(4w+\ell)\lambda \\ (4w+\ell-1)(4w+\ell)\lambda - r(4w+\ell) \end{bmatrix}$$

Now observe that every entry on the last row of the matrix on the L.H.S. is divisible by 8. So every entry of the last row of matrix on the R.H.S. is also divisible by 8 , i.e., $(4w + \ell)[\lambda(4w + \ell - 1) - r] \equiv 0 \pmod 8$.

(i) Let $\ell = 2$. Then $4w + 2 \equiv \pmod 4$ and hence $\lambda - r \equiv 0 \pmod 4$.

(ii) Let $\ell = 1$. Then $(4w + 1)[4w\lambda - r] \equiv 0 \pmod 8$, i.e., 8 divides $4w\lambda - r$. So r must be even and Lemma 10.15(ii) implies λ is even. Hence $4w\lambda \equiv 0 \pmod 8$ and therefore $r \equiv 0 \pmod 8$.

(iii) Let $\ell = -1$. Then 8 divides $\lambda(4w - 2) - r$ and hence r is even. Again Lemma 10.15 (ii) implies λ is even. Therefore 8 divides $4\lambda w$ and hence 8 divides $2\lambda + r$. The following result of Calderbank [40] can now be obtained.

Theorem 10.17. Let D be a 2-(v, k, λ) design with intersection numbers $s_1, s_2, \ldots, s_n$. If $s_1 \equiv s_2 \equiv \ldots \equiv s_n \equiv 0 \pmod 2$, then either

(i) $r \equiv \lambda \pmod 4$

or

(ii) $k \equiv 0 \pmod 4$ and $v \equiv \pm 1 \pmod 8$.

The proof of Theorem 10.17 uses all the previous lemmas and Result 10.14 in particular. Using similar ideas, Calderbank [40] also proves.

Theorem 10.18. Let D be a 2-(v, k, λ) design with intersection numbers $s_1, s_2, \ldots, s_n$. If $s_1 \equiv s_2 \equiv \ldots \equiv s_n \equiv 1 \pmod 2$, then either

(i) $r \equiv \lambda \pmod 4$

or

(ii) $k \equiv v \pmod 4$ and $v \equiv \pm 1 \pmod 8$.

Using the above two theorems, Calderbank [40] immediately rules out designs no. 4, 5, and 19 from Neumaier's table. We state this result as

Corollary 10.19. The following q.s. 2-(v, k, λ) designs with intersection numbers x, y do not exist:

(i) $2\text{-}(21, 9, 12);\ x = 3, y = 5.$
(ii) $2\text{-}(21, 8, 14);\ x = 2, y = 4.$
(iii) $2\text{-}(21, 7, 3);\ x = 1, y = 3.$

With considerably more effort, Calderbank [40] also rules out designs no. 1, 12, and 18 from Neumaier's table.

Theorem 10.20. The following q.s. 2-(v, k, λ) designs with intersection numbers x, y do not exist:

(i) $2\text{-}(19, 9, 16);\ x = 3, y = 5.$
(ii) $2\text{-}(24, 8, 7);\ x = 2, y = 4.$
(iii) $2\text{-}(33, 9, 6);\ x = 1, y = 3.$

The case where the intersection numbers of a 2-(v, k, λ) design are all congruent modulo an odd prime was considered in another recent paper by Calderbank [41]. We state his result below and refer to [41] for the proof.

Theorem 10.21. Let p be an odd prime and D be a 2-(v, k, λ) design with intersection numbers $s_1, s_2, \ldots, s_n$ satisfying $s_1 \equiv s_2 \equiv \ldots \equiv s_n \equiv s \pmod{p}$. Then either

(i) $r \equiv \lambda \pmod{p^2}$,

(ii) $v \equiv 0 \pmod{2}$, $v \equiv k \equiv s \equiv 0 \pmod{p}$, $(-1)^{v/2}$ is a square in GF(p),

(iii) $v \equiv 1 \pmod{2}$, $v \equiv k \equiv s \not\equiv 0 \pmod{p}$, $(-1)^{(v-1)/2}\, s$ is a square in GF(p),

(iv) $r \equiv \lambda \equiv 0 \pmod{p}$ and either

(a) $v \equiv 0 \pmod{2}$, $v \equiv k \equiv s \not\equiv 0 \pmod{p}$,

(b) $v \equiv 0 \pmod{2}$, $k \equiv s \not\equiv 0 \pmod{p}$, v/s is a non-square in GF(p),

(c) $v \equiv 1 \pmod{2p}$, $r \equiv 0 \pmod{p^2}$, $k \equiv s \not\equiv 0 \pmod{p}$,

(d) $v \equiv p \pmod{2p}$, $k \equiv s \equiv 0 \pmod{p}$,

(e) $v \equiv 1 \pmod{2}$, $k \equiv s \equiv 0 \pmod{p}$, v is a non-square in GF(p),

(f) $v \equiv 1 \pmod{2}$, $k \equiv s \equiv 0 \pmod{p}$, v and $(-1)^{(v-1)/2}$ are squares in GF(p).

As an application of Calderbank's results, this rules out some q.s. designs whose existence was not known in M.S. Shrikhande [146].

Corollary 10.22. The following q.s. (v, b, r, k, λ) designs D with intersection numbers x, y do not exist:

(i) 2 - (21, 105, 65, 13, 39); $x = 7, y = 9$.

(ii) 2 - (41, 205, 85, 17, 34); $x = 5, y = 8$.

(iii) 2 - (43, 301, 175, 25, 100); $x = 13, y = 16$.

Proof For (i), apply Theorem 10.17. For (ii) and (iii) use Theorem 10.21 with $p = 3$.

We can also rule out the existence of the first two exceptional parameters from Baartmans and M.S. Shrikhande [6] (see Chapter VIII), by applying Theorem 10.21 with $p = 3$ and 23, respectively. This result appears in M.S. Shrikhande [153].

Corollary 10.23. The following q.s. (v, b, r, k, λ) designs with intersection numbers x, y do not exist.

(i) $2 - (232, 638, 99, 36, 15); x = 0, y = 6.$
(ii) $2 - (5290, 13202, 861, 345, 56); x = 0, y = 23.$

Concluding Remarks 10.24. We have attempted to show in this chapter how coding theory has recently provided some new tools that are valuable in design theory. We mention also the recent papers of Calderbank [38], [39] which make use of the so-called linear programming method of Delsarte [60]. The paper [38] was referred earlier in Chapter IX. We cite M.S. Shrikhande [153] for a recent survey on q.s. designs, intersection numbers, and codes. See also Jungnickel and Tonchev [99].

REFERENCES

[1] R.W. Ahrens and G. Szekeres, On a combinatorial generalization of 27 lines associated with a cubic surface, J. Austr. Math. Soc. 10, 485-492, (1969).

[2] M. Aschbacher, On collineation groups of symmetric designs, J. Comb.Theory (A) 11, 272-281, (1971).

[3] E.F. Assmus, Jr. and J.H. van Lint, Ovals in symmetric designs, J. Comb. Theory (A) 27, 307-324, (1977).

[4] A. Baartmans, K. Danhof, and R. Holliday, Parallelism in quasi-symmetric designs, Ars Combin. 16, 77-84, (1984).

[5] A. Baartmans, K. Danhof, and S. Tan, Quasi-residual quasi-symmetric designs, Discrete Math. 30, 69-81, (1980).

[6] A. Baartmans and M.S. Shrikhande, Designs with no three mutually disjoint blocks, Discrete Math. 40, 129-139, (1982).

[7] A. Baartmans and M.S. Shrikhande, A characterization of the extensions of symmetric designs, Discrete Math. 57, 301-305, (1985).

[8] B. Bagchi, No extendable biplane of order 9, J. Comb. Theory (A), 49, 1-12, (1988).

[9] B. Bagchi, Personal communication , (1989).

[10] E. Bannai, On tight designs, Quart. J. Math. (Oxford) 28, 433-448, (1977).

[11] E. Bannai and T. Ito, Algebraic Combinatorics I: Association Schemes, Benjamin Cummings, Menlo Park, Calif., 1984.

[12] L.M. Batten, Combinatorics of Finite Geometries, Cambridge University Press, Cambridge, 1986.

[13] H. Beker and W. Haemers, 2-designs having an intersection number k-n, J. Comb. Theory (A) 28, 64-81, (1980).

[14] E.R. Berlekamp, Algebraic Coding Theory, McGraw Hill, New York, 1968.

[15] T. Beth, D. Jungnickel and H. Lenz, Design Theory, B. I. Wissenschaftverlag, Mannheim, 1985 and Cambridge University Press, Cambridge, 1986.

[16] N.L. Biggs, Finite Groups of Automorphisms, London Math. Soc. Lecture Note Series 6, Cambridge Univ. Press, 1971.

[17] N.L. Biggs, Algebraic Graph Theory, Cambridge University Press, Cambridge, 1974.

[18] N.L. Biggs and A.T. White, Permutation Groups and Combinatorial Structures, London Math. Soc. Lecture Note Series 33, Cambridge University Press, 1979.

[19] D. Billington, A simple proof that all 1-designs exist, Discrete Math. 42, 321-322, (1982).

[20] R.C. Bose, On the construction of balanced incomplete block designs, Ann. Eugenics 9, 353-399, (1939).

[21] R.C. Bose, Strongly regular graphs, partial geometries, and partially balanced designs, Pacific J. Math. 13, 389-419, (1963).

[22] R.C. Bose, W.G. Bridges, and M.S. Shrikhande, A characterization of partial geometric designs, Discrete Math. 16, 1-7, (1976).

[23] R.C. Bose, W.G. Bridges, and M.S. Shrikhande, Partial geometric designs and two-class partially balanced designs, Discrete Math. 21, 97-101, (1978).

[24] R.C. Bose and W.S. Connor, Combinatorial properties of group divisible incomplete block designs, Ann. Math. Statist. 23, 367-383, (1952).

[25] R.C. Bose and D.M. Mesner, On linear associative algebras corresponding to association schemes of partially balanced designs, Ann. Math. Statist. 30, 21-38, (1959).

[26] R.C. Bose and K.R. Nair, Partially balanced incomplete block designs, Sankhyā 4, 337-372, (1939).

[27] R.C. Bose and T. Shimamoto, Classification and analysis of partially balanced incomplete block designs with two associate classes, J. Amer. Statist. Assoc. 47, 151-184, (1952).

[28] R.C. Bose and M.S. Shrikhande, On a class of partially balanced incomplete block designs, J. Statist. Plann. Inference 3, 91-99, (1979).

[29] R.C. Bose and S.S. Shrikhande, Geometric and pseudo-geometric graphs $(q^2+1, q+1, 1)$, J. Geom. 2, 75-94, (1972).

[30] R.C. Bose, S.S. Shrikhande, and N.M. Singhi, Edge regular multigraphs and partial geometric designs with an application to the embedding of quasi-residual designs, in: Teorie Combinatorie, Toma I, Atti dei Convenge Lincei. 17, Acad. Naz. Lincei, Rome, 49-81, 1976.

[31] A. Bremner, A diophantine equation arising from tight 4-designs, Osaka J. Math. 16, 353-356, (1979).

[32] W.G. Bridges and M.S. Shrikhande, Special partially balanced incomplete block designs and associated graphs, Discrete Math. 9, 1-18, (1974).

[33] A. Brouwer and J.H. van Lint, Strongly regular graphs and partial geometries, in: Enumeration and Design- Proc. Silver Jubilee Conf. on Combinatorics, Waterloo, 1982 (D.M. Jackson & S.A. Vanstone, eds.), Academic Press, Toronto,1984, 85-122.

[34] A. Brouwer, A.M. Cohen and A. Neumaier, Distance Regular Graphs, Springer Verlag , Berlin, Heidelberg and New York, 1989.

[35] R.A. Brualdi, Introductory Combinatorics, North Holland, New York, Oxford, Amsterdam, 1977.

[36] R.H. Bruck, Finite nets II: Uniqueness and embedding, Pacific . J. Math. 13, 421-457, (1963).

[37] R.H. Bruck and H.J. Ryser, The non-existence of certain finite projective planes, Canad. J. Math. 1, 88-93, (1949).

[38] A.R. Calderbank, Symmetric designs as the solution of an extremal problem in combinatorial set theory, Europ. J. Combinatorics, 8, 171-173, (1987).

[39] A.R. Calderbank, Inequalities for quasi-symmetric designs, J. Comb. Theory (A), 48 (1), 53-64, (1988).

[40] A.R. Calderbank, The application of invariant theory to the existence of quasi-symmetric designs, J. Comb. Theory (A), 44, 94-109, (1987).

[41] A.R. Calderbank, Geometric invariants for quasi-symmetric designs, J. Comb. Theory (A), 47, 101-110, (1988).

[42] A.R. Calderbank and P. Morton, Quasi-symmetric 3-designs and elliptic curves, SIAM J. Discrete Math. (to appear).

[43] P.J. Cameron, Extending symmetric designs, J. Comb. Theory (A), 14, 215-220, (1973).

[44] P.J. Cameron, Near regularity conditions for designs, Geom. Dedicata 2, 213-223, (1973).

[45] P.J. Cameron, Biplanes, Math. Z., 131, 85-101, (1973).

[46] P.J. Cameron, Quasi-symmetric designs possessing a spread, Proceedings of the 1988 Combinatorics Conference in Ravello, (to appear).

[47] P.J. Cameron, Locally symmetric designs, Geom. Dedicata 3, 65-76, (1974).

[48] P.J. Cameron and D.A. Drake, Partial $\lambda-$ geometries of small nexus, in: Combinatorial Math., Optimal Designs and their Appl., Proc. Fort Collins 1978, Ann. Discrete Math. 6 (J. Srivastava, ed.), North-Holland, Amsterdam, 1980, 19-29.

[49] P.J. Cameron and J.H. van Lint, Graphs, Codes, and Designs, London Math. Soc. Lecture Note Series 43, Cambridge Univ. Press, 1980.

[50] P.J. Cameron, J.M. Goethals, and J.J. Seidel, Strongly regular graphs having strongly regular subconstituents, J. Algebra 55, 257-280, (1978).

[51] P.J. Cameron, Parallelisms of Complete Designs, London Math. Soc. Lecture Note Series 23, Cambridge Univ. Press, 1976.

[52] R.D. Carmichael, Introduction to the Theory of Groups of Finite Order, Boston, 1937 (reprint Dover- New York 1956).

[53] L.A. Carmony and S. Tan, S-quasi-symmetry, Discrete Math. 34, 17-24, (1981).

[54] L.C. Chang, The uniqueness and the non-uniqueness of the triangular association scheme, Sci. Record Peking Math. (New Series) 3, 604-613, (1959).

[55] S. Chowla and H.J. Ryser, Combinatorial problems, Canad. J. Math. 2, 93-99, (1950).

[56] W.S. Connor, The uniqueness of the triangular association scheme, Ann. Math. Statist. 29, 262-266, (1958).

[57] G.M. Constantine, Combinatorial Theory and Statistical Design, John Wiley and Sons, New York, 1987.

[58] J.H. Conway and V. Pless, On the enumeration of self-dual codes, J. Comb. Theory (A) 28, 26-53, (1980).

[59] D.M. Cvētkovic, M. Doob, and H. Sachs, Spectra of Graphs, Academic Press, New York, 1980.

[60] P. Delsarte, An algebraic approach to the association schemes of coding theory, Philips Research Reports Supplements No. 10, (1973).

[61] P. Dembowski, Finite Geometries, Springer-Verlag, Berlin 1968.

[62] J. Dēnes and A.D. Keedwell, Latin Squares and their Applications, Academic Press, New York-London, 1974.

[63] R.H.F. Denniston, Some new 5-designs, Bull. London Math. Soc. 8, 263- 267, (1976).

[64] H. Enomoto, N. Ito, and R. Noda, Tight 4-designs, Osaka J. Math. 16, 39-43, (1979).

[65] R.A. Fisher, An examination of the different possible solutions of a problem in incomplete block designs, Ann. Eugenics 10, 52-75, (1940).

[66] A.M. Gleason, Weight polynomials of self-dual codes and the MacWilliams identities, in: Actes Congr. Inter. Math. Nice 1970, Vol. 3, 211-215, Gauthier-Villars, Paris 1971.

[67] J.M. Goethals and J.J. Seidel, Strongly regular graphs derived from combinatorial designs, Canad. J. Math. 22, 597-614, (1970).

[68] W. Haemers, A generalization of the Higman-Sims technique, Indag. Math. 40, 445-447, (1978).

[69] W. Haemers, Eigenvalue methods, in: Mathematical Centre Tracts 106, (1979) " Packing and Covering in Combinatorics" (A. Schrijver, ed.), 15-38.

[70] W. Haemers, A non-existence result for quasi-symmetric designs, in: Bose Memorial Conference, Indian Statistical Institute, Calcutta (to appear).

[71] W. Haemers, Regular two-graphs and extensions of partial geometries, (preprint).

[72] W. Haemers, A new partial geometry constructed from the Hoffman- Singleton graph, in: Finite geometries and designs, London Math. Soc. Lecture Series 49, Cambridge Univ. Press 1981, 119-127.

[73] W. Haemers and D.G. Higman, Strongly regular graphs with strongly regular decompositions, Linear Algebra and Appl., 114/115, 379-398, (1989).

[74] M. Hall, Jr., Combinatorial Theory, Second Edition, John Wiley and Sons, New York, (1986).

[75] M. Hall, Jr. and W.S. Connor, An embedding theorem for balanced incomplete block designs, Canad. J. Math. 6, 35-41, (1954).

[76] H. Hanani, Balanced incomplete block designs and related designs, Discrete Math. 11, 255-369, (1975).

[77] F. Harary, Graph Theory, Addison-Wesley, Reading, Mass., 1969.

[78] A. Hedayat and S. Kageyama, The family of t-designs - Part I, J. Statist. Plann. Inference 4, 173-212, (1980).

[79] M.D. Hestenes and D.G. Higman, Rank 3 groups and strongly regular graphs, in: Computers in Algebra and Number Theory, (G. Birkhoff and M. Hall, Jr., eds.) SIAM-AMS Proc. IV, Amer. Math. Soc. Providence, R.I., 1971, 141-159.

[80] D.G. Higman, Finite permutation groups of rank 3, Math. Z., 86, 145 156, (1964).

[81] D.G. Higman, Strongly regular designs and coherent configuration of type $\begin{pmatrix} 3 & 2 \\ & 3 \end{pmatrix}$, preprint.

[82] D.G. Higman, Coherent Algebras, Linear Algebra and Appl. 93, 209-239, (1987).

[83] D.G. Higman and C.C. Sims, A simple group of order 44,352,000, Math. Z., 105,110-113, (1969).

[84] S.A. Hobart, Designs of type (2, 2; 4), Ph.D. thesis, Univ. of Michigan, 1987.

[85] S.A. Hobart, A characterization of t-designs in terms of the inner distribution, (preprint).

[86] A.J. Hoffman, On the polynomial of a graph, Amer. Math. Monthly, 70, 30-36, (1963).

[87] A.J. Hoffman, On the exceptional case in the characterization of the arcs of a complete graph, IBM J. Res. Develop. 4, 487-497, (1960).

[88] A.J. Hoffman, On the uniqueness of the triangular association scheme, Ann. Math. Statist. 31, 492-497, (1960).

[89] A.J. Hoffman and R.R. Singleton, On Moore Graphs with diameters 2 and 3, IBM J. Res. Develop. 4, 497-504, (1960).

[90] R. Holliday, Quasi-symmetric designs with $y = \lambda$, Congressus Numerantium 48, 195-201, (1985).

[91] X. Hubaut, Strongly regular graphs, Discrete Math. 13, 357-381, (1975).

[92] D.R. Hughes, On t-designs and groups, Amer. J. Math. 87, 761-778, (1965).

[93] D.R. Hughes and F.C. Piper, On resolutions and Bose's Theorem, Geom. Dedicata 5, 129-133, (1976).

[94] D.R. Hughes, Semi-symmetric 3-designs, in: Finite Geometries(eds. N.L. Johnson,M.J. Kallaher, C.T. Long), Marcel Dekker, New York- Basel 1983, 223-235.

[95] D.R. Hughes and F.C. Piper, Design Theory, Cambridge University Press, Cambridge, (1986).

[96] D.R. Hughes, Extended partial geometries: dual 2-designs, (preprint).

[97] Q.M. Hussain, On the totality of the solutions for the symmetrical incomplete block designs: $\lambda= 2$, $k=5$ or 6, Sankhya 7, 204-208, (1945).

[98] N. Ito, On tight 4-designs, Osaka J. Math. 12, 493-522, (1975) (Corrections and Supplements, Osaka J. Math. 15, 693-697, (1978)).

[99] D. Jungnickel and V.D. Tonchev, On symmetric and quasi-symmetric designs with symmetric difference property and their codes, J. Comb. Theory (A) (to appear).

[100] S. Kageyama and A. Hedayat, The family of t-designs - Part II, J. Statist. Plann. Inference 7, 257-287, (1983).

[101] C.W.H. Lam, L. Thiel, S. Swiercz and J. McKay, The non-existence of ovals in the projective plane of order 10, Discrete Math, 45, 319-322, (1983).

[102] C.W.H. Lam, L. Thiel and S. Swiecz, The non-existence of finite projective planes of order 10, Canad. J. Math. XLI, 6, 1117-1123, (1989).

[103] E.S. Lander, Symmetric Designs: An Algebraic Approach, London Math. Soc. Lecture Note Series, 74, Cambridge Univ. Press, 1983.

[104] H. Lenz and D. Jungnickel, On a class of symmetric designs, Arch. Math.33,590-592,(1979).

[105] N.B. Limaye, S.S. Sane, and M.S. Shrikhande, The structure of triangle-free quasi-symmetric designs, Discrete Math. 64, 199-207, (1987).

[106] H. Lüneburg, Transitive Erweiterungen endlicher Permutationsgruppen, Lecture Notes in Mathematics 10, Springer, Berlin- Heidelberg-New York, 1965.

[107] F.J. MacWilliams and N.J.A. Sloane, The Theory of Error Correcting Codes, North-Holland, Amsterdam, 1977.

[108] K.N. Majumdar, On some theorems in combinatorics relating to incomplete block designs, Ann. Math. Statist. 24, 379-389, (1953).

[109] M. Marcus and H. Minc, A Survey of Matrix Theory and Matrix Inequalities, Allyn and Bacon, Boston, 1964.

[110] V.C. Mavron and M.S. Shrikhande, Designs with intersection numbers 0 and 2, Arch. Math. 52, 407-412, (1989).

[111] T.P. McDonough and V.C. Mavron, Symmetric designs and geometroids, Combinatorica 9 (1), 51-57, (1989).

[112] D.M. Mesner, An investigation of certain combinatorial properties of partially balanced incomplete block experimental designs and association schemes, with a detailed study of designs

of Latin square and related types, unpublished doctoral thesis, Michigan State Univ. (1956).

[113] A. Meyerowitz, Quasi-symmetric designs with $y = \lambda$, (preprint).

[114] A. Meyerowitz, S.S. Sane, and M.S. Shrikhande, New results on quasi-symmetric designs - an application of MACSYMA, J. Comb. Theory (A), 43, 277-290, (1986).

[115] L.J. Mordell, Diophantine Equations, Academic Press, London-New York, 1969.

[116] A. Neumaier, Regular cliques in graphs and special 1 1/2 designs, in: Finite Geometries and Designs, London Math. Soc. Lecture Note Series 49, Cambridge University Press, 1981, 244-259.

[117] A. Neumaier, Regular Sets and quasi-symmetric 2-designs, in: Combinatorial Theory (D. Jungnickel and K. Vedder, eds.), Lecture Notes in Math., 969, (1982), 258-275.

[118] R.M. Pawale, Quasi-symmetric 3-designs with triangle-free graph, Geom. Dedicata, (to appear).

[119] R.M. Pawale and S.S. Sane, A short proof of a conjecture on quasi-symmetric 3- designs, Discrete Math., (to appear).

[120] R.M. Pawale, Inequalities and bounds for quasi-symmetric 3-designs, (J. Comb. Theory (A), (to appear).

[121] S.E. Payne and J.A. Thas, Finite Generalized Quadrangles, Pitman Press, Boston, 1984

[122] C. Peterson, On tight 6- designs, Osaka J. Math. 14, 417-435, (1977).

[123] W.W. Peterson and E.J. Weldon, Error-Correcting Codes, Second edition, MIT Press, 1972.

[124] V. Pless, Introduction to the Theory of Error Correcting Codes, John Wiley and Sons, New York, 1982.

[125] V. Pless and N.J.A. Sloane, On the classification and enumeration of self-dual codes, J. Comb. Theory (A) 18, 313-335, (1975).

[126] A. Pott and M.S. Shrikhande, A note on t-designs with few intersection numbers, Discrete Math., (to appear).

[127] D. Raghavarao, Constructions and Combinatorial Problems in Design of Experiments, John Wiley, New York, 1971 (reprint Dover Paperback Series 1989).

[128] D. Raghavarao and M.S. Shrikhande, A generalization of special partially balanced designs to higher associate class designs, Ars Combin. 6, 97-111, (1978).

[129] D.K. Ray-Chaudhuri and R.M. Wilson, On t-designs, Osaka J. Math. 12, 737-744, (1975).

[130] N.N. Roghelia and S.S. Sane , Classification of (16, 6, 2)-designs by ovals, Discrete Math. 51,167-177, (1984).

[131] G.C. Rota, Editor, Studies in Combinatorics, The Mathematical Association of America, 1978.

[132] H.J. Ryser, Combinatorial Mathematics, Wiley, New York, 1963.

[133] S.S. Sane, On extendable planes of order 10, J. Comb. Theory (A) 38, 91-93, (1985).

[134] S.S. Sane, Hussain chains revisited, Discrete Math. 70, 211-213, (1988).

[135] S.S. Sane, Lecture Notes on Witt designs, prepared for the summer school in combinatorics and graph theory, Bangalore, 1984.

[136] S.S. Sane, On a family of symmetric designs, in: Proceedings of the Seminar on Combinatorics and its Applications in honor of S.S. Shrikhande, (K.S. Vijayan & N.M. Singhi, eds.), Indian Statistical Institute, Calcutta, 1982, 292-302.

[137] S.S. Sane, Affine subplanes of order 3 in projective planes of order 4 and quasi-multiple designs, Mitt. Sem. Univ. Giessen, 155, (1985).

[138] S.S. Sane and M.S. Shrikhande, Finiteness questions in quasi-symmetric designs, J. Comb. Theory (A) 42, 252-258, (1986).

[139] S.S. Sane and M.S. Shrikhande, Quasi-symmetric 2, 3, 4-designs, Combinatorica, 7 (3), 291-301, (1987).

[140] S.S. Sane and M.S. Shrikhande, Quasi-symmetric designs and biplanes of characteristic three, J. Comb. Theory (A), (to appear).

[141] S.S. Sane, S.S. Shrikhande and N.M. Singhi, Maximal arcs in designs, Graphs and Combinatorics 1, 97-106, (1985).

[142] M.P. Schutzenberger, A non-existence theorem for an infinite family of symmetrical block designs, Ann. Eugenics, 14, 286-287, (1949).

[143] J.J. Seidel, Strongly regular graphs with (-1, 1, 0) adjacency matrix having eigenvalue 3, Linear Algebra and Appl., 1, 281-298, (1968).

[144] J.J. Seidel, Strongly regular graphs, in: Surveys in Combinatorics, (B. Bollabos, Ed.), London Math. Soc. Lecture Note Series, 38, Cambridge Univ. Press, 1979, 157-180.

[145] M.S. Shrikhande, Strongly regular graphs and quasi-symmetric designs, Utilitas Math. 5, 297-309, (1973).

[146] M.S. Shrikhande, Strongly regular graphs and quasi-symmetric designs II, Unpublished, (1973).

[147] M.S. Shrikhande, On a class of negative latin square graphs, Utilitas Math, 5, 293-303, (1974).

[148] M.S. Shrikhande, Strongly regular graphs and group divisible designs, Pacific J. Math. 54, 199-207, (1974).

[149] M.S. Shrikhande, Designs with triangle-free graph, in: Proceedings of the Seminar on Combinatorics and its

Applications in honor of S. S. Shrikhande, (K.S. Vijayan & N.M. Singhi, eds.) Indian Statistical Institute, Calcutta, 1982, 334-339.

[150] M.S. Shrikhande, A survey of some problems in combinatorial designs-a matrix approach, Linear Algebra and Appl. 79, 215-247, (1986).

[151] M.S. Shrikhande, Quasi-symmetric designs and quadratic polynomials, in: Colloquia Mathematica Societatis Janos Bolyai 52, Combinatorics, Eger (Hungary), 1987, 471-480.

[152] M.S. Shrikhande, Recent results on designs and their intersection numbers, in: Proceedings of the Symposium on Optimization, Design of Experiments, and Graph Theory, Indian Institute of Technology, Bombay, Dec. 15-17, 1986, 77-89.

[153] M.S. Shrikhande, Designs, intersection numbers, and codes, in: IMA Volume in Mathematics and its Applications , (D.K. Ray-Chaudhuri, ed.), Vol. 21, Coding Theory and Design Theory, Part II: Design Theory, 1990, 304-318.

[154] M.S. Shrikhande, On the parameters of a certain exceptional block design, Utilitas Math. 26, 103-108, (1984).

[155] M.S. Shrikhande and N.M. Singhi, An elementary derivation of the annhilator polynomial of extremal (2s+l) - designs, Discrete Math. 90, 93-96, (1990).

[156] M.S. Shrikhande and N.M. Singhi, s-quasi-symmetric designs, Congressus Numerantium 36, 69-76, (1982).

[157] S.S. Shrikhande, The impossibility of certain symmetrical balanced incomplete block designs, Ann. Math. Statist. 21, 106-111, (1950).

[158] S.S. Shrikhande, On the dual of some balanced incomplete block designs, Biometrics 8, 66-72, (1952).

[159] S.S. Shrikhande, On a characterization of the triangular association scheme, Ann. Math. Statist. 30, 39-47, (1959).

[160] S.S. Shrikhande, The uniqueness of the L_2 association scheme, Ann. Math. Statist. 30, 781-798, (1959).

[161] S.S. Shrikhande, Strongly regular graphs and symmetric 3-designs, in: A Survey of Combinatorial Theory, (J.N. Srivastava, et. al., eds.), North Holland, New York, 1973, 404-409.

[162] S.S. Shrikhande and Bhagwandas, Duals of incomplete block designs, Journal of Indian Statistical Assoc. 3, 30-37, (1965).

[163] S.S. Shrikhande and D. Raghavarao, Affine α-resolvable incomplete block designs, in: Contributions to Statistics, Pergamon Press, Oxford, 1964, 471-480.

[164] S.S. Shrikhande and N.M. Singhi, Adjacency multigraphs and embeddings : A survey, in: Combinatorics and Graph Theory, Proceedings, Calcutta 1970, (S.B. Rao, ed.), Lecture Notes in Math. 885, 111-132, 1980.

[165] C.C. Sims, On the isomorphism of two groups of order 44,352,000, in: The theory of finite groups (R. Brauer & C.H. Sah, eds.), Benjamin, 1969, 101-108.

[166] R.G. Stanton and J.G. Kalbfleisch, Quasi-symmetric balanced incomplete block designs, J. Comb. Theory (A) 4, 391-396, (1968).

[167] A.P. Street and D.J. Street, Combinatorics of Experimental Design, Oxford Univ. Press, Oxford, 1987.

[168] L. Teirlinck, Non-trivial t-designs without repeated blocks exist for all t, Discrete Math, 65, 301-311, (1985).

[169] J.A. Thas, On 4-gonal configurations, Geom. Dedicata 2, 317-326, (1973).

[170] J.A. Thas, Construction of partial geometries, Simon Stevin 46, 95-98, (1973).

[171] J.A. Thas, Constructions of maximal arcs and partial geometries, Geom. Dedicata, 3, 61-74, (1974).

[172] V.D. Tonchev, Quasi-symmetric designs and self-dual codes, Europ. J. Combinatorics, 7, 67-73, (1986).

[173] V.D. Tonchev, Embedding of the Witt-Mathieu system S(3, 6, 22) in a symmetric 2-(78, 22, 6) design, Geom. Dedicata 22, 49-75, (1987).

[174] V.D. Tonchev, Quasi-symmetric 2-(31, 7, 7) designs and a revision of Hamada's conjecture, J. Comb. Theory (A), 42, 104-110, (1986).

[175] V.D. Tonchev, Combinatorial Configurations, Pitman Monographs and Surveys in Pure and Applied Mathematics 40, John Wiley and Sons, New York, 1988.

[176] J.H. van Lint, On ovals in PG (2, 4) and the McLaughlin graph, in: Papers dedicated to J. J. Seidel, (edited by P. J. Doelder, J. de Graaf and J. H. van Lint), Eindhoven Univ. of Technology Press, 1984, 234-254.

[177] W.D. Wallis, Combinatorial Designs, Marcel Decker, New York, 1989.

[178] H. Wielandt, Finite Permutation Groups, Academic Press, New York-London, 1964.

[179] H.A. Wilbrink and A.E. Brouwer, A (57, 14, 1, 4) strongly regular graph does not exist, Math. Centrum Preprint, ZW 121/78, 1978.

[180] R.M. Wilson, On the theory of t-designs, in: Enumeration and Design, Academic Press, New York, 1984, 19-49.

[181] R.M. Wilson, An existence theory of pairwise balanced designs, II. The structure of PBD closed sets and the existence conjectures, J. Comb. Theory (A), 13, 246-273, (1972).

[182] E. Witt, Über Steinersche Systeme, Abh. Math. Sem. Hamburg 12, 265-275, (1938).

INDEX

adjacency matrix 17
affine
 -design 35
 -plane 8
 -space 13
arc 132
 -(α, n) arc 101
 -maximal arc 132
association scheme 20, 49, 60
automorphism 23, 30
auxillary set of matrices 126

Baer subplane 102
balanced incomplete
 block design (BIBD) 4
biplane 45
block 1
 -design 4
 -graph 36
 -residual 129
Bose-Connor property 50, 63
Bose's method of differences 14
Bruck-Ryser-Chowla
 Theorem 8, 125

Cameron's Theorem 15
Cauchy interlacing 83
conference
 -matrix 92
 -graph 93, 94
code
 -binary 166, 193
 -doubly even 194
 -dual 192, 193
 -Golay 99
 -self dual 193
 -self orthogonal 193
 -ternary 16
complementary
 -design 38
 -graph 20
complete
 -design 1
 -graph K_n 21
 -m-partite graph K(m,n) 21
critical q.s. design 156
Cvetcovic bound 89

decomposition
 -graph 87
 -exceptional 93
design 1
 -block 4
 -complete 1
 -critical 166
 -derived 9
 -dual 6
 -exceptional q.s. 159
 -extendable 9
 -group divisible 62
 -Hadamard 11
 -isomorphic 1
 -Paley 11
 -quasi-affine 127
 quasi-residual 8
 -quasi-symmetric (q.s.) 34
 -residual 8

-resolvable 13
-symmetric 6
-t-(v, k, λ) 1
-tight 178
-triangle-free q.s. 155
-trivial 1
-Witt 35, 99
dodecads 117

eigenvalue 18
eigenvector 18
exceptional
parameters 47, 48, 144
-q.s. design 141, 142
extendable 9
extension of a
symmetric design 16, 41

Fano plane 2, 11
Fisher's inequality 6
flags 4
i-flats 12
Frobenius' Theorem 19

Gaussian coefficient 12
generalized quadrangle 33
generator matrix 193
geometric 31
Golay code 99
graph
-block 37
-Clebsch 22
-cocktail party 42
-complete bipartite
K(m, n) 21
-complete K_n 21
-connected 26, 77
-geometric 31
-Gerwitz 106
-Higman-Sims 52, 100, 107
-Hoffman-Singleton 100, 119
-ladder 42
-lattice $L_2(n)$ 22
-line 97
-m-partite 21
-McLaughlin 121
-negative latin square $NL_g(n)$ 56
-Paley 11, 25, 28
-Petersen 21
-pseudo-geometric 31
-rank 3 24
-regular 17
-Schlafli 64
-strongly regular (S.R.) 17, 20
-triangular T(n) 22
group
-orbits 23
-dihedral 30
-Mathieu 99
-permutation 17
-rank 3 24
-transitive 22
group divisible design (GDD) 62

Hadamard
-2-design 11
-3-design 2
-matrix 11
Hall-Connor Theorem 8, 46
Hall's conjecture 152
Hamada's conjecture 192, 198
Hasse-Minkowski invariant 7

Higman-Sims graph 52, 100
Hoffman coclique bound 89
Hoffman polynomial 20, 81
Hoffman-Singleton graph 119
Hoffman's Theorem 18
hyperplanes 12
hyperoval 101

incidence matrix 1
intersection numbers 34
interlacing 83
irreducible matrix 17

latin square 56, 97
lattice graph $L_2(n)$ 22
line
- -passant 101
- -secant 101
- -tangent 101

linear programming method 206

Mathieu groups 99, 100
matrix
- -adjacency 17
- -auxillary 118
- -conference 92
- -generator 193
- -Hadamard 2, 11
- -incidence 1

multiplicities of eigenvalues 19
negative latin square graph
$NL_g(n)$ 56
net 31, 32
orbit 22, 23
order 23
orthonormal basis 18
oval 101

Paley
- -design 11
- -graph 25

parallel class 13, 32
parallelism 163, 164
parameters 1
partial geometric design 80
partial geometry 30
partial λ-geometry 81
partially balanced incomplete
block design (PBIBD) 20, 49, 62
permutation group
- -regular 22
- -transitive 22

Perron's Theorem 18
plane
- -affine 8
- -projective 8

primitive decomposition 87
projective space 12

quadric 64
quasi-affine design 127
quasi-symmetric (q.s.) 34
- -2-design 34
- -3-design 56
- -4-design 46

rank
- -of a group 23
- -of a matrix 6

rationality conditions 27, 38
Ray-Chaudhuri and
Wilson Theorem 176
regular
-graph 17
-decomposition 87
regular set (in a graph) 143
residual 8
resolvable 13

s-quasi-symmetric 47
self orthogonal 193
semi-regular group divisible
design (SRGDD) 63
semi-symmetric design 81
special partially balanced
incomplete block design
(SPBIBD) 49, 73
Singhi's conjecture 152
spread 164
Steiner system $S_1(t, k, v)$ 4
strongly regular design 85
strongly regular (SR) graph 20, 51
symmetric
-design 6
-group S_n 24

t-design 1
Teirlinck's result 4
tight designs 178
threshold symmetric
design 125
transversal design 97
triangle-free q.s. design 108, 130
triangular graph T(n) 22
two-way counting 2, 3, 35

Wilson's result 4

Witt designs 99

Printed in the United States
By Bookmasters